Communicat

IN SEARCH OF MEDIA

Götz Bachmann, Timon Beyes, Mercedes Bunz,
and Wendy Hui Kyong Chun, Series Editors

Pattern Discrimination
Markets
Communication
Machine
Remain

Communication

Paula Bialski, Finn Brunton,
and Mercedes Bunz

IN SEARCH OF MEDIA

University of Minnesota Press
Minneapolis
London

meson press

In Search of Media is a collaboration between the
University of Minnesota Press and meson press,
an open access publisher, https://meson.press/.

Published by the
University of Minnesota Press, 2019
111 Third Avenue South, Suite 290
Minneapolis, MN 55401-2520
https://www.upress.umn.edu

in collaboration with
meson press
Salzstrasse 1
21335 Lüneburg, Germany
https://meson.press

ISBN 978-1-5179-0647-4 (pb)
A Cataloging-in-Publication record for this book is available
from the Library of Congress.

Contents

Series Foreword

"Media determine our situation," Friedrich Kittler infamously wrote in his Introduction to *Gramophone, Film, Typewriter.* Although this dictum is certainly extreme—and media archaeology has been critiqued for being overly dramatic and focused on technological developments—it propels us to keep thinking about media as setting the terms for which we live, socialize, communicate, organize, do scholarship, et cetera. After all, as Kittler continued in his opening statement almost thirty years ago, our situation, "in spite or because" of media, "deserves a description." What, then, are the terms—the limits, the conditions, the periods, the relations, the phrases—of media? And, what is the relationship between these terms and determination? This book series, *In Search of Media,* answers these questions by investigating the often elliptical "terms of media" under which users operate. That is, rather than produce a series of explanatory keyword-based texts to describe media practices, the goal is to understand the conditions (the "terms") under which media is produced, as well as the ways in which media impacts and changes these terms.

Clearly, the rise of search engines has fostered the proliferation and predominance of keywords and terms. At the same time, it has changed the very nature of keywords, since now any word and pattern can become "key." Even further, it has transformed the very process of learning, since search presumes that, (a) with the right phrase, any question can be answered and (b) that the answers lie within the database. The truth, in other words, is "in there." The impact of search/media on knowledge, however, goes beyond search engines. Increasingly, disciplines—from sociology to economics, from the arts to literature—are in search of media as a way to revitalize their methods and objects of study. Our current media situation therefore seems to imply a new term, understood as temporal shifts of mediatic conditioning. Most broadly, then, this series asks: What are the terms or conditions of knowledge itself?

To answer this question, each book features interventions by two (or more) authors, whose approach to a term—to begin with: *communication, pattern discrimination, markets, remain, machine*— diverge and converge in surprising ways. By pairing up scholars from North America and Europe, this series also advances media theory by obviating the proverbial "ten year gap" that exists across language barriers due to the vagaries of translation and local academic customs. The series aims to provoke new descriptions, prescriptions, and hypotheses—to rethink and reimagine what media can and must do.

Machine Communication

**Paula Bialski, Finn Brunton,
and Mercedes Bunz**

This book searches for an understanding of communication, in
light of the fact that more communication than ever before is being
mediated digitally by machines. To understand the full scope of
what "to communicate" now means, it will curiously explore the
complexity of the entities we are communicating to, with, and
through to other entities. Looking not just at how we communicate
with digital media but also at how digital devices and software
communicate with us, to us, and to each other can more precisely
outline the power (imagined or not) that computers and the people
who take part in building our computers hold. By looking at various
dimensions of communication in history and practice, this volume
serves as an account of how digital media addresses its "subjects";
how alien and invisible the mediators we built have become; and
how complex communication is now that we work with and interact
with our machines. With this, the volume *Communication* takes up
the main theme of the book series In Search of Media: it searches
for the shift in the mediatic and technological conditioning of
communication and aims to make this shift visible.

Digital media are not just filters but are "vehicles that carry and
communicate meaning" (Peters 2012, 2). Because media carry,

relay, and sort information, they have the ability to meddle in our communication. Mediating is about meddling—and who meddles, what meddles, and how it/they meddle is key to understanding how digital communication functions. If communication does not unfold anymore between merely two (or more) conscious entities, yet rather includes an invisible third party, this can drastically shift what *to communicate* actually means. What (a set of programs, networking systems, or interfaces) or who (a team of developers) meddles in our communication is a crucial question for the technical realities of our societies today.

To understand the many modes and facets of this shift of communication, this book analyzes the communication of machines, experts, and aliens and turns to historic and contemporary engineers, designers, and users that are all taking part in how we humanly and nonhumanly communicate. For this, the volume's chapters look at machine communication, although from three different perspectives: in chapter 1, Finn Brunton explores communication and digital technology by showing the alienlike dialogue between technical entities; in chapter 2, Mercedes Bunz looks at how digital technology, which now has even started to speak, is addressing us; and in chapter 3, Paula Bialski studies machine communication when turning to the social aspects of technical systems. Or, in other words, chapter 1 looks at the nonhuman communication between computers and the indirect communication happening in digital infrastructure, chapter 2 looks at how computers communicate to humans and examines the force of communication, and chapter 3 looks at the actual creation of digital infrastructure and machine communication through a code review system.

With these three perspectives, the volume is bringing together three contrasting scholars: Brunton is a media theorist who has written about a multitude of media-related, historically rooted topics, including surveillance and obfuscation, as well as a cultural history of spam, in which he showed that it is not humans who are producing the majority of communication traffic. Bunz is a media and technologies scholar researching digital technology and philosophy of technology who has published on artificial intelligence, the internet of things, and algorithms, always questioning how those technical applications transform knowledge and, with it, questions of power. Bialski's background sits between sociology and ethnography of media. In the past years, she conducted ethnography around the way new media fosters new forms of mobility and togetherness, how it transforms our understanding of space (location) and intimacy. In 2016, she started a fieldwork project in a large-scale mapping software company in Berlin. Her chapter in this volume therefore draws on her ongoing ethnographic project with these corporate software developers.

Introducing the chapters in more detail, one could say that in the first chapter, Brunton looks at the ways we communicate, directly and indirectly, with digital infrastructure. Pushing the analysis of the inhuman aspect of this infrastructure further, he turns to aspects of digital communications history, starting with early interaction design and discussions by J. C. R. Licklider and Robert Taylor about whether two tape recorders communicate when they play to each other. Brunton's inquiry into nonhuman communication then brings up the problem of timekeeping in networks, ending with Google's own timekeeping system, TrueTime, all of them showing that communication means much *more* than just sending and receiving—in a complex setup, they produce the contemporary *now*. Tracing inhuman communication further, Brunton points also to the problem of automated trolling, as most of the communication that is sent and received is inhuman anyway. Following the hypothesis that we are in a process of building deeply inhuman architectures and systems on a vast scale, Brunton finally flips his approach by turning to historic projects of extraterrestrial communication in analogy to our current situation, aiming to open dimensions of analysis that might otherwise escape us.

Not far from Brunton's approach, the second chapter, authored by Bunz, also traces nonhuman communication by turning to the force that unfolds in digital communication. Like Brunton, Bunz is shifting the perspective on communication away from an anthropocentric or anthropomorphic approach. Starting with the observation that a certain *force* has always been a theme in theories of communication, she aims to identify the particular aspect of this force for digital technology by asking, How is digital technology addressing us? When studying communication, Bunz thus mirrors Brunton's approach, although she is turning it in the other direction: instead of looking toward and into digital communication systems, she looks at how digital communication systems are approaching us, thereby drawing on Althusser's theory of interpellation. Turning to the historic events in digital design, such as the introduction of Apple's iPad in 2010, by analyzing digital brand communication reliance on little animals as mascots and by bringing to the fore the shift of historical storytelling through Google's Doodles, she shows that digital interfaces are addressing us as very young children. This has not necessarily to be read negatively, as it also calls on experimental–operational knowledge, which can be traced to the early history of graphical user interfaces and the influence of child psychologist Jean Piaget on computer scientists, especially on Seymour Papert and Alan Kay. Like Brunton's, Bunz's chapter also then turns in a very different direction: after following the question after the force of digital communication to its paradigm of infantilization, she becomes interested in the paradigm itself and how it refrains from following a well-behaved dialectical thinking typical for the nonhuman logic of technology—it is manipulating us at the same time as it is empowering us.

Finally, Bialski's chapter offers a rich ethnographic case study that, much like Bunz and Brunton do, explores communication with as well as through technology. After spending nearly two years at a large corporate software company in Berlin, she looks at software developers at work—specifically the way they review one another's lines of code through a standardized, mandatory "code review system." This system, while being mechanic and seemingly mundane, is also a highly variable communicative process because of the culture of communication that develops around it. Here she shows how technical systems emerge out of both human and machine communication. Through her chapter, she draws on examples of software developers at work—communicating with one another and with their machines, and waiting (and relying) on their machines to "communicate" with other machines. Through this, she analyzes how a technical system structures cooperation and how standards of communication develop. By looking at the idiosyncrasies of human–human as well as human–machine communication, she aims to provide a grounded example of the multifaceted nature of communication in digital cultures.

What unifies all three approaches in this volume is that all chapters aim to show that there has been a shift in our communication toward an interaction *with* or *among* machines, which comes across more strongly as the three approaches cover very separate ground. Brunton carefully underlines the "opportunity to consider how we engage with machines and how machines engage with each other"; Bunz explores how "machines seem to engage with *us*"; and Bialski shows that there is a communicative interrelationship between the compilers, databases, processors, memory, servers, "clouds," and their programmers, through the infrastructure within which both the programmers and the machines function. By ethnographically, historically, and theoretically exploring the nonhuman part in communication, by turning to machine communication, this small volume hopes to contribute to existing theories of communication.

Reference

Peters, John Durham. 1999. *Speaking into the Air: A History of the Idea of Communication*. Chicago: University of Chicago Press.

[1]

Hello from Earth

Finn Brunton

Human beings write a great deal about the essence of matter. It
would be nice for matter to begin to write about the human mind.
—Lichtenberg

Do Two Tape Recorders Communicate?

J. C. R. Licklider and Robert Taylor coauthored a landmark paper in
1968: "The Computer as a Communication Device." It was a major
public step in shifting the understanding of what computers are *for*:
from massive specialized calculators to communications platforms
for the interaction of many users augmented by computation. It
is part of the same cultural and technological moment as Douglas
Engelbart's *Mother of All Demos,* a theatrical happening (part technology showcase, part live science fiction film) at the 1968 Fall Joint
Computer Conference that displayed mice, outliners, real-time document collaboration and word processing, the nuances of linked
documents, and other hallmarks of interactive and interpersonal
computing (Bardini 2000, 138–42). (It's part of the same long event
of inventing interaction that included Kay, Papert, and Piaget, so
elegantly chronicled and analyzed by my colleague Mercedes Bunz
in chapter 2 of this book.) Like Engelbart's showcase, Licklider and

2   Taylor's article was only partially about the technology itself—about the metal, glass, and code: it was also a vehicle for communicating a feeling.

With a background in experimental psychology, Licklider was always conscious of the human in the newly emerging loop that Engelbart (1962, 2) called the "whole system" of people and machines. Licklider (1988, 30) documented his thought process, folding himself in as an experimental subject, a test pilot, one of the first people to be sitting at the computer console "four or five hours a day." He wrote about the feeling of hitting the "brain–desk barrier," assimilating the new assembly of information pulled together and presented by this networked library-computer (Licklider 1965, 102). He talked about the "motivational trap" of a cunningly built interactive terminal, drawing you deep into the structure of a problem or a concept (Greenberger 1962, 208). "This is going to revolutionize how people think," he repeated (Licklider 1988, 29). Not individual people, either, but groups, institutions, communities: "a flow of metal and ideas and of flexibility and change," he wrote, in a single phrase that captures the poetic thrill of engineering not a particular technology but the overarching "system system" (Licklider 1963, 628). Forget processing payroll or running Navier–Stokes equations: this was a change to human *work,* in the deepest sense, both individually and collectively. Engelbart called it "augmentation"; Licklider called it "symbiosis." "The Computer as a Communication Device" was a path from new communication to new forms of *community,* sharing with "all the members of all the communities the programs and data resources of the entire super community" (Licklider and Taylor 1968, 32).

To make this case, in 1968, Licklider and Taylor first had to clarify a term: "A communications engineer thinks of communicating as transferring information from one point to another in codes and signals." That's not what they meant by "communication." They were trying to get at something else: "our emphasis on people is deliberate" (21).

"To communicate," they wrote, "is more than to send and to receive. Do two tape recorders communicate when they play to each
other and record from each other? Not really—not in our sense. We
believe that communicators have to do something nontrivial with
the information they send and receive" (Licklider and Taylor 1968,
21). They talked about "the richness of living information." About
being "active participants" whose "minds interact," about "creative
aspects" that "transcend" the transmission of information. What
this turned out to mean for them, in practice, is using computers
to produce models that people can manipulate in real time over
the network. By "communication," they meant the comparison of
mental models.

This may not seem like much after all the talk of transcendent
interacting minds, but Licklider and Taylor were dealing with the
assumptions of a particular audience schooled in Claude Shannon's
information theory, where "communication" can indeed be defined
down to the transmission of information between senders and
receivers over channels. By emphasizing the role of interacting
humans in the use of computers, they were taking up a novel and
potent idea. Licklider and Taylor were trying to counteract the
model of communication-as-information-transmission to make a
case for time-sharing and better user interfaces. They drew strate-
gically on the deep resonant legacy of the word *communication.* But
where does this powerful resonance come from—and what does it
actually mean to communicate?

Shannon, approaching it as a telecommunications engineering
problem, simplified communication to *information*: any exchange
can be understood quantitatively in terms of entropy and probability, in transmissions between senders and receivers over variably
noisy channels, and thus the coding, compression, and capacity
of channels can be designed appropriately. Warren Weaver,
who introduced, popularized, and expanded on Shannon's work,
wrote in the introduction to the landmark *Mathematical Theory of*

Communication, "The word *communication* will be used here in a very broad sense to include all of the procedures by which one mind may affect another. This, of course, involves not only written and oral speech, but also music, the pictorial arts, the theater, the ballet, and in fact all human behavior" (Shannon and Weaver 1949, 3). Weaver was making the classic cybernetic rhetorical move, listing a whole family of seemingly disparate things all linked together by the power of the theory: all can be understood informationally, starting from a minimal state with no particular content beyond the probability of any given bit.

If we squint a little, this approach is similar to very different areas of media and communications studies (each answering the question, as Bunz puts it, of the "force" at work in any act of communication beyond what is conveyed). McLuhan, amid all his Joycean hubbub, proposed a model of media that is not dissimilar: the actual content of a book, a movie, a TV show, is more or less a *distraction* from the medium itself, which is what really communicates. The medium shapes linear print minds who sort the world into taxonomic ontologies, and it shapes "cool" tactile global villagers who listen into the acoustic space of broadcast for distant cultural thunder. The content is not what we should be reading if we want to understand what's going on. McLuhan was the mentor and inspiration to USCO, a new media arts collective devoted to evoking the new consciousness theoretically made available through the immersive experience of electronic media. They built light installations and optical meditation machines and environments like *The World,* a colossal project in an airplane hangar in Garden City with eighteen slide projectors controlled by a repurposed IBM mainframe, 16mm film projection, and cutting-edge real-time analog video (Kuo 2008, 136). McLuhan spent some time in person with USCO, and Gerd Stern and Michael Callahan once had to drive him to the airport from the University of Rochester, in October 1964. McLuhan was getting into one of those now-rare prop planes with a stairwell in the tail. "I remember," recalls Callahan, "Marshall walking up the stairs and us standing below and seeing him disappear into the plane. But then we saw his legs, his feet came back a few steps, and he leaned out and said, 'Disregard the content and concentrate upon the effect'" (Kuo 2008, 133).

Disregard the content: in McLuhan's analysis, to get at what is happening in the act of communication, we can deliberately ignore whatever is being communicated, which is generally a mere epiphenomenon of the *event* of a particular medium, whether movable type or TV. His famously odd remark about the electric light—that it's a medium, the only medium, that has form but no content—makes sense in the context of his argument: electric light carries no information of its own, beyond off and on, but its presence changes how we live and how we think, "the change of scale or pace or pattern that it introduces," in ways that escape our observation precisely because we focus on what the light falls on, or what the newspaper article is about (McLuhan 1994, 8).

Similarly, in a more nuanced approach, JoAnne Yates (1989) and Cornelia Vismann (2008) understand how communications systems—particularly the tools of bureaucracy and management, files, forms, and paperwork—interpellate people, calling them to account and putting them into all kinds of arrangements and subject relationships. It's not just the policeman Bunz describes, hailing us in the street, but also the citation he issues and the paperwork we must fill out. The preset fields, systems of identifiers, and needle-sort card files *order* us, in both senses of the word, and do so in their structure and not just their written content in ways that exceed the kind of interpretative tools we've developed. Process generated, the universe of memos, forms, circulars, manuals, reports, and tables assembled into "control through communication" constituted a coordination mechanism that used words in enormous volume to increase efficiency across firms. The whole tool kit developed for reading and interpreting written media—hermeneutic approaches, textual analysis, and criticism—falters in the face of the most quantitatively significant written modes that shape our lives: bureaucratic and legal documents and the boxes, cabinets, shelves, and pallets of paperwork.

Vismann draws on Friedrich Kittler, who most purely exemplifies this strategy of getting to the heart of "communication" and "media" by subtracting what naively seems to be the most important part—the subjective, interpersonal content, the experience of communicating. Kittler (1990, 370) builds directly on his reading of Shannon's work, with the humans on either end of the line merely distractions, atavistic holdouts whose fixation on meanings and inner experience keeps us from seeing the operation of the system of communication itself:

> An elementary datum is the fact that literature (whatever else it might mean to readers) processes, stores, and transmits data, and that such operations in the age-old medium of the alphabet have the same technical positivity as they do in computers. . . . What remains to be distinguished, therefore, are not emotional dispositions but systems.

That parenthetical, with its casual dismissal of most of what we might think literature involves, has the chilling mildness of HAL 9000 locking the human out of the ship in *2001: A Space Odyssey.* Kittler's move in the media theory of his time was not simply to apply ideas from Shannon, Turing, and sound and image reproduction technologies to get new insights into media as such but to read *backward* from these developments. It's not that communications media are now different. They have always "determine[d] our situation," in his famous phrase, from flowing humanist–Romantic handwriting and magic lanterns to ancient Greek music and mathematics (Kittler 1999, xxxix). (In this, he inverts McLuhan's other key tenet—that media are "the extensions of man," expanding and externalizing our bodies and senses—by making humans the extensions of media.) Depersonalized mechanical and electronic media just make that determining power more obvious.

If we accept Kittler's (2010) argument, then Weaver is not wrong to write that the mathematical theory of communication can apply to "in fact all human behavior." With that theory in mind, Kittler asserts, we can give a better (truer, more accurate, more useful) account of media than, for instance, "literary scholars" with a "trivial, content-based approach to media" (31). "Let us therefore," Kittler writes, "forget humans, language, and sense in order to move on to the particulars of Shannon's five elements and functions instead" (44). (It must be said here that Shannon himself, a humble and clear-eyed scientist, distrusted the hyping of his theory, even within the scientific and engineering community: "It has perhaps ballooned to an importance beyond its actual accomplishments," he wrote in a 1956 editorial. "Seldom do more than a few of nature's secrets give way at one time" [3].)

Kittler read Shannon carefully and came out with a model arguing that we miss the main event of communication, the actual operation of the network of discourse, if we are distracted from close attention to the storage and transmission of data by people's alleged inner lives and subjectivity. Licklider, who also read Shannon very carefully (as well as knowing him personally), came to precisely the opposite conclusion: "A communications engineer thinks of communicating as transferring information from one point to another in codes and signals," he wrote with Taylor. The danger lay in having too shallow, too *depersonalized,* an understanding of communication. Cutting the humans out of the loop was a design problem, producing models of computer networking that couldn't enrich human thinking. Licklider was getting at what Engelbart, in 1962, called communication technologies as "a way of life":

> We refer to a way of life in an integrated domain where hunches, cut-and-try, intangibles, and the human "feel for a situation" usefully coexist with powerful concepts, streamlined technology and notation, sophisticated methods, and high-powered electronic aids. (1)

"Do two tape recorders communicate when they play to each other and record from each other?" No, said Licklider and Taylor: the communicators have to do something with the information they send and receive—something *nontrivial,* they write, in the mathematical sense of a trivial proof being easy, obvious, productive of a shallow truth.

Again, not to make too much of semantics, but notice the clashing "trivialities" here: for Kittler, the content-based, human-centric approach is a "trivial" one, whereas for Licklider and Taylor, the mere transmission of signals is the trivial part; the "nontrivial" happens somewhere between the interface, the computer, the human mind, and the interpersonal community of collaborators.

What I want to argue for here is a third position that doesn't supersede the prior work but builds on it. It starts not from people nor from media in general but from *computing* in particular. In this, my argument supplements studies of digital formats and interfaces (e.g., Sterne 2012; Bardini 2000; Bunz, chapter 2), protocols (e.g., Galloway 2004; Bratton 2016), and infrastructures (e.g., Starosielski 2015; Edwards 2003; Sandvig 2013)—but it comes to the problem of understanding the effect of digital networks on communication by a different route and with a different destination. I will demonstrate that the concept of trivial or nontrivial communication has become much more ambiguous and complex to discuss in the thirty years since Kittler and the fifty years since Licklider and Taylor's foundational work. I will make the case for this position in three parts: first, by outlining the hybrid, social–technological complexity of the seemingly simple matter of keeping time among networked computers; second, by discussing the limited role played by humans in the communications taking place over the internet and associated technologies; and finally, by presenting the history of formats for communication with alien intelligence as offering useful analogies to our situation. In a mirror image of Bunz's essay, I assert that we can productively reimagine *what we are talking to* when communicating with digital systems and considering their communications with us. This argument does not push humans to the margin, after Kittler—as vacuous metabolic vehicles across which discourse networks transact, like the parasitic typewriters in a William Burroughs story—but neither does it put them at the center of the story, as Licklider and Taylor do. I will start by asking

Mutually Suspicious Clocks

Much of what structures contemporary networked life and digital media is "tape recorders" talking to each other, understood as various forms of addressable data storage, processing, and retrieval. The most active areas of current digital communication—analyzed qualitatively or quantitatively—are almost entirely tape recorders in networks of exchanges, for which the humans are only of third- or fourth-order consequence. (This is another way of phrasing the "layers" and stacks that Bunz describes in detail when we look at digital media in particular.) These are areas where the idea of "triviality" becomes very tricky.

Consider the problem of the clock.

The clock is the very model of dehumanized Machine Age operations, illustrated by the people who embraced it as the new model of work and life, like Alexei Gastev's ultra-Taylorist institutes incorporating clock rhythm into the cyclogram-trained, wordless operation of machine shop labor coordinated by the "electronic beeps of a machine" or Louis Aragon's perfect Modernist line: "In my left pocket I carry a remarkably accurate self-portrait: a watch in burnished steel. It speaks, marks time and understands none of it" (Stites 1989, 154; Ades 2006, 181). What could be more trivial than the working of a clock?

Licklider was one of the main motors, with Robert Fano and others at MIT, of the Compatible Time-Sharing System (CTSS), a multiuser computing project with remote terminals connected to a central mainframe. Licklider and Fano wanted computing to feel like an invisible resource, always on, a daily utility like electricity or water. (One of their great triumphs was the complaints they would receive whenever the system was down—why can't I get access right now? What the hell's going on?—because that meant the users were already taking it for granted as a quotidian faucet of computation [Lee and Rosin 1992, 26].) The tape recorders were there in CTSS, literally: banks of IBM 729 magnetic tape drives, which you can watch spinning back and forth through their address space over the shoulders of scientists being interviewed on television in the 1960s. Licklider, Fano, and the tape recorders were supporting the epitome of the symbiosis-and-augmentation computers-helping-humans-communicate model.

They had a timing problem. The machines didn't have clocks, so if you hit a glitch or a loop, you locked up the whole machine for everybody, and there was no mechanism that could time stamp files or kill a process after a few seconds. To resolve this, they figured out how to hook up a basic time-of-day clock to the machine through the printer port (Waldrop 2001, 234). This seems like the very definition of trivial communication: a simple clock to regulate all those dumb tape recorders, coordinating the most basic form of communication—a statement of the immediate present moment—to aid the nontrivial communication of time-sharing computer network users.

Jump forward to September 1985, to a document called the Request for Comments (RFC) 956, one in the series of coordinating memos for the architects and committees creating the internet and related technologies—the long-term result of CTSS and other experiments. "The recent interest within the Internet community in determining accurate time from a set of mutually suspicious network clocks has been prompted by several occasions in which gross errors were found in usually reliable, highly accurate clock servers after seasonal thunderstorms which disrupted their primary power supply" (Mills 1985, 1). Once you start networking the machines together, the problems of keeping time become far more complex. How do you know which messages come first? What's the order in the queue? How do you make sure data aren't being resent unnecessarily? (The software consultant Mathias Verraes [2015] captured it with a sardonic joke: "There are only two hard problems in distributed systems: 2. Exactly-once delivery 1. Guaranteed order of messages 2. Exactly-once delivery.")

It's one thing to have a master clock regulating all the activities, like the Taylorist tick of scientifically managed factory labor waiting for the shift whistle. This is the clock as understood by Lewis Mumford, the prototype of all other industrial machines and the coordinating center of the humans–tools–processes megamachine, building pyramids and cranking out Model Ts. Network clocks are a very different class of machine. They have to account for transmission lag and mechanical failures and the behavior of many other "mutually suspicious" clocks. The clock protocols have to compensate for the effects of seasons and weather on parts of the network.

The clock is now so big that it has to contain and compensate for Earth's atmosphere, because it interferes with the accuracy of the clock. John Durham Peters (2015) has discussed elemental forms as media; Jussi Parikka (2015) has argued for studying the geological histories embedded in and sometimes expressed by media systems. We don't even have to go as far as artworks or solar energy captured in trees as paper pulp to make their case with this example: lightning will hit high-tension lines in late summer thunderstorms and disrupt the overall temporal picture for the network, so tools are developed to correct for that. At this point, we're still below the level of triviality assigned to tape recorders that "play to each other and record from each other": these are still just the *clocks* that enabled the tape recorders to play back and forth in an orderly fashion thirty years ago.

As I write this, the current situation is far more complex. It's global, operating across diverse media from undersea fiber-optic cables to microwave relays, satellites, cellular-band radio, and copper twisted pair. It faces signal delays and lag and clock drift from machine to machine (which is affected by the amount of work the machine is doing, and even by heat). To work out the time, it will take a signal that has been time stamped in one location to reach its destination, and we rely on the Global Positioning System (GPS) (whose scale Bunz describes in the context of our overreliance on the voice of navigation instructions). GPS, being satellite based, has to compensate for relativistic problems—the way time passes differently outside of Earth's gravity well. The clock must indirectly incorporate not only the atmosphere but the shape of Earth itself as the "geoid," the subtle undulations of the gravitational field, expressible as a hypothetical ocean surface and perturbing satellites in and out of their true paths.

All of this work to define the machine's time on the network has to function in the context of geopolitics and political economy as expressed in one of their purest forms: time zones. Knowing the time goes well beyond signaling, measuring, and geodesy—and not just for the benefit of the human glancing at her phone's lock screen but for automatically time stamping and logging events, correlating them across borders, accurately representing the hour in the past and the future, and providing data that can account for the national solar time at any given place. Will people be at work? When should alarms be set? Will the stock market be open? Should lights be on or off? There are clock systems particular to computing, like UNIX time, used by many file formats and operating systems, incrementing one second per second since 00:00:00 UTC, January 1, 1970—at this exact moment, 1,456,328,317. ("UTC" is the Coordinated Universal Time standard, mean solar time at 0°, used in aviation, weather forecasting, many scientific applications, and the Network Time Protocol, keeping computer clocks in synch.) On top of these systems runs the stack of Westphalian assertions of identity and human chronotypes concerned with hunger, work, and sleep.

The time zone system can be read as an ongoing, minimalist history of modernity: territorial struggles, global trade, new technologies and forms of work, all embedded as brief lines in computer databases like the "tz" or "zoneinfo" database in UNIX-based systems (Internet Assigned Numbers Authority 2016). Samoa decided to move across the international dateline to be on the same day as countries to the east (Australia, New Zealand, China) and lost a day in 2011 doing so; those countries had become more important than the ships from San Francisco had once been. Nepal is deliberately fifteen minutes ahead of India; Indiana has a complex, politically fraught relationship with U.S. Daylight Savings Time (itself a matter of issues between agricultural, industrial, and white-collar labor) that puts some counties in the state an hour ahead of others. China went from five time zones to one in 1949, a potent tool for centralization and unity—all the way to the Xinjiang Region is theoretically on Beijing Standard, though Uyghurs also use and maintain their own time. Crimea's interim government switched to Moscow Standard Time in late March 2014, a gesture of political significance on par with Spain's adoption of Central European Time in 1940, chronometrically cementing Franco's alliance with Hitler (and stranding Spain, much of which lies to the west of Greenwich, in a time zone that creates its own daily rhythm of "late" sunrises and sunsets).

All of this and more must be automatically accounted for by the network's clocks, expressed as code with a laconic history of conflicts, alliances, powerful markets, and subaltern peoples. The clock has become a kind of political appliance, having to incorporate all this complex history to function properly:

```
Rule Libya     1997 only - Oct 4 0:00 0 -
Rule Libya     2013 only - Mar lastFri 1:00 1:00 S
Zone Europe/Simferopol 2:00 EU EE%sT 2014 Mar 30 2:00
4:00 - MSK 2014 Oct 26 2:00s
Zone Asia/Jerusalem 2:20:54 - LMT 1880 2:20:40 - JMT
1918 # Jerusalem Mean Time? 2:00 Zion I%sT
```

Consider our clock at this point. To coordinate interactions between "tape recorders" across the network—all those different points to and from which data are sent, requested, retrieved—it directly and indirectly incorporates millions of global clocks with their various skews and drifts, in the political governance patchwork of dozens of time zones on which both humans and clock-triggered events rely, in a network topology in which particular nodes can be congested or completely down and that need to be routed around, which changes the timing on transmissions, with times synchronized over radio and transoceanic cables sunk in pelagic sediment, requested from atomic clocks with reference to satellites and quartz crystals—themselves synched by human labor, with gatherings to monitor the "leap second" transition, including the influence of nothing less than space-time itself. The business of efficiently managing message queues and flows carries us through the step-by-step of pragmatic engineering decisions into the foundations of time and light.

We can go still one step further before returning to the question of the relative triviality of communications: into the *now* of this vast timekeeping process. One of Google's greatest achievements is a distributed database called "Spanner." Spanner "shards" data over hundreds of data centers across the planet, maintaining consistency on a global scale. Thus users can see the same text in a collaborative online document; search results can be appropriately ranked, mail sent and received, ads priced and displayed, planetwide and without internal contradictions (Corbett et al. 2012). Data repeatedly requested can be copied to a shard closer to the location of the request, to lower the latency, delivering the next request faster; something can happen on a phone in Kazakhstan, a server in Seoul, and a laptop in Cape Town, simultaneously (or as close to simultaneous as the human mind is capable of observing unassisted), and be reconciled and coordinated; failure of the system in one place is compensated for in another before any human intervention is necessary. To make this possible, Google's engineers were obligated to develop their own timekeeping system, TrueTime, which uses "time master machines" in each data center, cross-referencing GPS with their own atomic clocks; those latter units have the rather awe-inspiring name of "Armageddon masters." Even this has uncertainty for which it must compensate, though: TrueTime, constantly correcting itself to account for

drift, uncertainty, and slew, has a real "now" of between one and
fourteen milliseconds long.

It does not detract from the achievement of Spanner and the True-
Time system to point this out; I bring it up to make clear that there
is nothing simple, nothing *trivial* about the act of tracking time for
the tape recorders. That is precisely what is happening here: to
enable the coordination of the storage, transmission, and retrieval
of data, prior to and independent of any encounter with humans,
we have built a system that includes late summer storms, the
curvature of space-time, and a new, specialized, continuous now,
regulated by an arrangement called "Paxos leader lease protocol"
(to whom should we listen, all of us tape recorders, when updating
the state of our data?). It is a wonder of engineering not least
because it vanishes into the background—because we are unaware
of the labor of producing *now*. Jimena Canales (2009) has captured
the emergence of another new now at the turn of the nineteenth
century, a world on the threshold of a tenth of a second and the
influence of this new now in domains ranging from experimental
psychology and the exact sciences to semiotics and the birth of
photography and movies. It was a critical chapter in the story of
"measuring, mastering, and disenchanting" in the production of the
modern age (207). There was nothing trivial about either the tools
or the implications of pushing "now" past a decisecond. Likewise,
creating a simultaneous, global, networked now of between one
and fourteen milliseconds has implications that cut across media,
society, and applied science: it sets up the framework within which
what Wendy Chun (2008, 149) calls the "enduring ephemeral"
of "constantly disseminated and regenerated digital content" is
experienced, the cultural signature of our age.

In other words, Licklider and Taylor's title "The Computer as a
Communication Device" was perfectly accurate, but in the opposite
sense: the computer is a communication device, and the communicators are doing something nontrivial with the information they
send and receive. It's just that very, very few of the communicators
are *people*.

Telephones Have Conversations

After we have given so much attention to the problems of networked time and timing, of maintaining state and keeping data synchronized across the breadth of the network, it seems logical that we would, at last, turn to the humans who are producing the traffic. All that timekeeping—itself not really communicative, except in the most minimal sense—is the service, after all, of human–human and human–machine interactions. What of the "active participants" with "minds interacting" for whom this whole system was to serve as the platform? It is not mere contrarianism to say that quantitatively and—I argue—increasingly qualitatively, human activities are of less and less significance. In this, I seek not to contradict but to complement Bunz's essay, which describes the ambiguities (empowering or patronizing?) in how digital technologies talk to us as children. As actual communication on present networks moves away from anthropocentric models, it shifts the "situation" she describes and with it the meaning of those infantilizing forms of interface addressed to us humans. Her study of those modes of interface and address frees me to turn to the nonhuman side of the situation (to paraphrase Licklider and Taylor, my emphasis on machines is deliberate). I will approach this in three parts: that humans are not producing as much of the communicative traffic as we may think; that, in some areas, our activity is not easily distinguished from the activity of machines; and that our communicative activity often has its greatest effect in the aggregate, as data rather than as the expressions of individuals.

First, humans are not producing all of the traffic or even, in some cases, the majority. There's spam in volumes that dwarfs human–human email exchanges, very little of which human readers encounter: it is generated using templates with "per-message polymorphism" by botnets running on malware-infected computers, sent in million-message batches to automated filtering systems—that is, systems trying to identify human text in an arms race with systems that are not trying to *imitate* humans but to model what the filters think human behavior is (Brunton 2013, 182). On bad days on the network, upward of 80 percent of all the email sent is spam (again, only a tiny portion of which will ever meet human eyes). This is of a piece with anywhere on the network that human attention pools and aggregates.

Google, maintaining an index of links between sites as a picture of what people consider most important or notable for any given query, has to contend with bot-driven posting activity on social media, pirated wikis, and in the comments everywhere. "Great post, really interesting" writes a commenter, with "interesting" linking to a client's site or an affiliate scam or a porn-clip landing page. The system of Private Blog Networks has been built solely to establish the appearance of a thriving human community where none exists. Twitter and other social networks regularly have to conduct sweeps to purge the bot accounts from their ranks (bots that follow paying users in packs of thousands to make them look important and popular). The same is true of buying likes on Facebook or YouTube, listens on SoundCloud, clicks on ads, and so on. In a bankable version of Goodhart's law ("When a measure becomes a target, it ceases to be a good measure"—or, "What gets evaluated, gets gamed"), any metric meant to describe human interest, esteem, or attention more generally will produce purpose-built nonhumans who will take it over for pay (Goodhart 1981, 116). It's the most perverse version of Licklider's goal of "man–machine symbiosis," because the bots and the business models they represent don't want to overwhelm the system or ruin the arrangement. Sometimes, in some corners of the internet, on poorly secured fora where bots can set up accounts unchecked, you can experience a Philip K. Dick–like moment of existential vertigo: is *anyone* here human other than me?

The second point follows from the second clause of that question. "Other than me": it may not matter if I or my program is writing, as human and machine communication activities can be indistinguish-able in many domains. One of the most telling of the Snowden disclosures was the revelation of the research going into what is called "information-operations" work, "influence or disruption," in

18   projects like JTRIG in the United Kingdom and OPERATION EARNEST
VOICE in the United States (Greenwald 2014). The goal is to "Deny/
Disrupt/Degrade/Deceive" online conversations, communities, and
movements through the injection of massive numbers of generated "personas" that can engage in online social activity (Fielding and
Cobain 2011). They can shift sentiments, quietly spread propagan-
da, amplify messages, give the impression of trending or significant
topics, game online polls, or simply flood a conversation that would
otherwise be potentially significant with irrelevancies, noise, and
derailing fight starting.

These personas may not necessarily be artificial; they may or may
not be conversation engines running on a deep learning feature
set. There may indeed be people on the keyboard at the other
end—as in the case of Vladimir Putin's troll factory in St. Peters-
burg, with coordinated teams at work manipulating conversations
about Ukraine and Syria, or gamergate trolls using a mix of bots
and their own inimitable human capacity for bullying, harassment,
threats, and time-wasting interference to drive women out of
gaming and off the internet (Sindelar 2014). There are many
significant things to consider about the consequences for political
discourse and how we conceive of a public sphere that includes
bot-saturated social networks, but the salient point is that, for the
practical purpose of excluding messages or kicking bots off the
platform, it *doesn't matter* if a human or a machine is communicating. Software development seeks to make them functionally
indistinguishable. (Analyzing a study of bots on the side of Labour
in the United Kingdom revealed that rules identifying bots were
often counting particularly fervid human supporters [Bartlett
2017].) The ultimate goal still involves humans, however indirectly:
to polarize their political perspectives, to waste their time, to create
an illusion of majority belief, to bury something and keep them
from seeing it. The humans matter as an end result, but not as the
means. Whether in a messaging app, on the help desk, or attacking
an unpopular Twitter user, these systems are meant to erase the
distinctions between our potential interlocutors.

We can look, for instance, at Lilly Irani's work on Amazon's Mechanical Turk system. Amazon goes to great lengths to make dealing
with the human workers as much like interacting with artificial
intelligence as possible, in keeping with the slogan of "Artificial
Artificial Intelligence." Irani (2015) calls the broader phenomenon
the "digital microwork industry": humans to finesse all the bits of
affective analysis that pattern-matching AI still struggles with (an
ironic echo of the original human "computers," occupationally
producing calculations before their work was mechanized). Beyond
the implications for labor, justice, and the organization of society,
I want to emphasize how deliberately vague the human and
machine work is at the client's interface: provision some capacity
and dump modular tasks into it and let the system crunch away.
Who knows if there's a sentient consciousness at work? Does it
matter? Seen in a certain light, John Searle's Chinese Room looks
like a knowledge worker's factory floor.

The gap keeps narrowing between social network bots (and many
other kinds of automated work, but this stays with our theme)
and what Anab Jain (2014) calls "meatpuppets." This is her term
for people who join an internet discussion solely to influence it,
and more broadly for how we as communicants on the network
can be understood: as livestock for producing information,
propagating memes, responding to automated notifications, and
generally being a felt glove with glued-on eyes on the limb of a
vast and intricate machine. A pack of adolescents activated by
text messages or forum posts to attack an individual or disrupt an
online conversation are, for purposes of managing the trouble they
create, just bots that have easier time passing the various automated tests that try to prove humanness. (This speaks to Bunz's point
that it's more important to look at the *situation* of communication
than the subjects involved—because who or what the subjects are
may be far more uncertain than we'd think.) Meatpuppets carry out
tasks and narrow the human–machine gap—like interns at content
farms, snorting Adderall and cranking out "content" in response to
click rates, competing with software that does precisely the same

thing with some occasional textual bumpiness but no need for office space, bathroom breaks, or amphetamine.

This meatpuppet condition is perfectly expressed by the internal structure revealed in the Ashley Madison hack. Ashley Madison, a dating site for married people theoretically looking for affairs, was hacked in purported retaliation for its failure to fully delete the data of users who paid to be erased from the service. What is significant, for our purposes, is what was disclosed when the hackers ("the Impact Team") dumped user records, corporate memoranda, and much more onto the public web for scrutiny. There were almost no actual women on the site. The business model was to draw men into loops of interacting with simple chatbot programs, some based on profiles mass created by spammers, to keep them hooked on paying to send messages, "virtual gifts," "winks," and the rest of the pixel tat common to dating and social apps. Internal documents refer to this fembot population as the "fraud-to-engager tool," turning merely curious male users into customers with a credit card on file (Newitz 2015). (Within the company, the bots were called "Angels.") It is not simply that the men were interacting with procedural systems mistaken for people but that the men were themselves behaving procedurally and reliably as simple stimulus-response, A/B-tested systems that may show complexity in person but in aggregate can be tapped for sap like a stand of maple trees.

As the consequences of the hacked data dump played out, even the blackmail was mostly automated: clever extortionists set up sites where the curious (spouses, coworkers, the general public) could type in a name to see if someone was present in the Ashley Madison database as a customer—which would trigger an email to the name's email address, threatening to reveal all the person's Ashley Madison activities to his social network contacts unless blackmail was paid (in Bitcoin, no less) (Krebs 2015). A user of Ashley Madison could have a complete, personal, Dreiserian tragedy play out—curiosity, temptation, flirting, offers and promises, exposure, shame, and secret extortion—that never once involved direct communication with another human being.

This brings me to the third point in my argument: the utility and significance for many purposes of aggregate data over individual expression on the network. The human contributions are often more useful in volume, as fodder for deep learning systems, predictive text algorithms, and translation engines, than as the outcome of any individual activity: not as much for what they tell us about you as for what they tell us about people like and unlike you. McKenzie Wark has discussed the unexpected side of the collapse of privacy on the network: not just the terrifying, targeted, Orwellian violations of personal autonomy for small numbers of dissidents, activists, and minorities worldwide—which was to be expected, tragically—but the *indifference* with regard to everyone else as individual identities (Gregg 2013). Your individual identity is in many ways the least interesting thing about you for purposes from advertising to social segmentation to assessing credit ratings to targeting a particular mobile phone signal for a drone strike: the ability to *address* and target you and others like and unlike you is more significant than your evaluation of what about you matters.

The common, lazy defense of surveillance—if you have nothing to hide, you have nothing to fear—is a useful illustration here because of how it misses the point: if you have nothing to hide, you are much more useful to a surveilling adversary because you help to distinguish those who *do* have something to hide. Your visibility probably does not matter (no one cares) but is helpful in how it informs the analysis of others. Your good credit aids in the identification of potential bad credit risks for discriminatory pricing. Individuals are merely biography; groups and social graphs are demography and Big Data, something far more useful. In other words, even when we are communicating with other humans ("directly," as mediated and managed by the network), the content of our communication is of consequence well beyond any partic- ular *meaning* it may have for us, as Big Data—data at a scale that calls into question any personal self-evaluation of their significance.

These aggregate data produce rapid improvements in the experience of communication with machines. Human–machine communication is becoming a rich, coherent, ubiquitous part of everyday life, exemplified by the proliferation of "conversational user interface," where all three of the points made so far in this section can be discerned.

"Saying that cultural objects have value," wrote Brian Eno (1996, 81) with a crisply aphoristic turn of phrase, "is like saying that telephones have conversations." Of course we have conversations over the telephone, *using* the telephone, with other people, as we transact value across cultural objects; but the implicit logic has begun to break down. Telephones, particularly mobile devices like smartphones, have conversations, and we have conversations with them. Phone and interphone conversations happen constantly— handshake check-ins about location, transmission of carrier information, Wi-Fi requests, and pushing and pulling background information—though that stretches the definition of "conversation" too far. We also have more direct and apparent conversations with telephones and with the systems doing voice recognition on telephones; think of Siri, Cortana, and customer service.

As digital platform interaction moves from computers to mobile devices, the trend is clear: many systems converge on one interface, and that interface is texting. It mediates interactions with services through the same messaging modes we think of as texting with humans, making it a perfect platform for increasing human– machine engagement, side-stepping the challenges of extracting human voices from ambient noise and accounting for accents and the diversity of speech. (As it happens, Licklider's career began in psychoacoustics, working on the problem of verbal communi- cation in a mechanically noisy environment like the cockpit of a plane under fire.) From fitness (Lark) to personal finance (Digit) to personal assistants (Magic) to games (Lifeline) to logistics (Taobao's 阿里小蜜) to news (Quartz) to payment (many) and customer service (Rhombus and more), whole categories of media activity that would once have implied custom software platforms, tools, or programs instead work through messaging. In those environments, the handoffs between humans and machines can be seamless. An exchange that triggers an automated response and one that pops up as a thread on a person's desktop or device, eliciting his response, are indistinguishable. As the most practical, quotidian matter, the easy binary split between people and machines that comes out in our prepositions—that we have conversations *with* people *over* telephones—is blurring.

All this contemporary activity may seem like rather old news to readers who recall Turing's (1950) rebuttals to the "arguments from various disabilities" against the possibility of artificial intelligence in his landmark paper "Computing Machinery and Intelligence." Turing's argument in favor of the possibility of computers having minds can be crudely summarized: if you think the machines are dumb, you should see the *people.* The argument builds on the assumption that machines can't be thinking because thinking is a special property of humanness, to which Turing responds, well, in what ways do you evaluate whether other people—or indeed you yourself—are thinking? Take that set of criteria and apply it to everything else, and see what passes; this is a way of eliminating what Reza Negarestani (2015) has called the "straw machine argument." Hence the Imitation Game, also known as the Turing test, which we pass in limited forms in the field constantly now—and here we touch closely on Bunz's half of this book, with the modes of address that are coming to typify that exchange. People give their credit card numbers away to seductive interaction designs; the programs that need to distinguish human users from all kinds of bots are in difficult straits; and even a living, breathing human user can effectively be a bot, or functionally indistinguishable from one. This boundary promises to become still more permeable now that so much of the research initiative is moving toward abstracting the most subjective, intimate areas of communication, from facial expressions to sentiment analysis to "interestingness."

In light of this, I argue that we are in the process of building deeply inhuman (which I do not mean in a pejorative sense) architectures and systems on a vast scale, whose content we partially constitute, and it is in the *context* of those systems that we now communicate—"the computer as a communication device." "In thought and political analysis, we still have not cut off the head of the king," Foucault (1990, 89) famously wrote; I'm worried that I still habitually think the king has a head. We did not make humanlike interlocutors, just humanlike interaction and interface designs that sit atop an infrastructure that has nothing whatsoever to do with the anatomy, physiology, or even cognitive processes that define the shifting borders of humanness. Then, as Bunz describes, we made it talk to us as if we were children, complete with cute animals. It is this infrastructure that we communicate through and, more and more often, *with*. We have theories about how we communicate with nonhumans of all kinds, from Lucy Suchman's situated actions to Donna Haraway's companion species to anthropological studies of humans in natural environments to theories of biosemiotics. I wonder if we could push them further: to think about exchanges with things that are alien to us in a different sense.

As it happens, another area of human imagination and scientific research is concerned with logical formalisms, binary pulses, and the communication of language and images to a "missing subject." It closely tracks and in some cases overlaps with the history of electromagnetic and computational communications media, and it shares a great deal with them. I believe it offers a useful analogy for our current situation: a series of examples of extraterrestrial media formats in which we as humans have attempted to communicate with the truly alien, in experimental arrangements we are now in the process of re-creating on our own planet.

Hello from Earth

Consider this thought experiment: we want to communicate with a potentially habitable planet. Therefore we send a signal that will take decades or centuries to reach a distant star, with no prior understanding of the biology or sensory anatomy of our communicants, much less their symbol systems or their technical instruments or the nature of their cognition. What to send? The work of communicating with aliens is the work of communicating with an entity that is for present purposes almost entirely unknowable—an entity that the anthropologist Klara Capova (2013), in her study of extraterrestrial signal work, calls the "missing subject." What, then, do we communicate?

This is a useful question to ask for two reasons.

The first is that the process of trying to develop formats, rules, systems, and messages for communicating with other entities that are fundamentally unknowable turns out to be very closely connected with many historical and contemporary problems in computing and telecommunications. The theoretical problem of communicating with aliens turns out to share a great deal with the practical problem of communicating with computers.

The second reason cuts a bit deeper: studying formats for extraterrestrial communication is a useful starting point because it necessarily forces a reevaluation of our human biases—bodily assumptions, cognitive habits, arrangements of language and what constitutes meaningfulness—that shape our concepts of what communication is. We have to carefully consider our assumptions about the other elements of the situation of communication that Bunz breaks apart in looking at how we are hailed and addressed. My hope is that these formats open dimensions of analysis that might otherwise escape us, which will apply to understanding communication over and with networked computers.

I focus on the act of communicating—broadcasting radio waves, flashing lights—precisely because it is the most demanding case. The Search for Extraterrestrial Intelligence (SETI) is usually divided into so-called passive and active SETI, where passive is listening: the fields of radiotelescopes scanning the skies for anything out of the ordinary. It is the search for potentially meaningful events, like the "Wow!" transmission from 1977—named for the remark Jerry

Ehman wrote on the printout next to the code representing the intensity of the signal (Gray and Ellingsen 2002). It's a fascinating and important process, but more pertinent are the projects in which we try to figure out how to *address ourselves* to our unknowable interlocutors—when we don't just look for regular or unusual patterns that could be meaningful to us but try to create *formats* that will be meaningful to an unknown and unknowable subject.

Mirrors and Morse Code:
A Clear Suggestion of Number and Order

Early attempts at an "interstellar language" are abstract more in an artistic than a mathematical sense, reading like proposals for vast Land Art initiatives and minimal sculptures on an enormous scale. Their components (mirrors, trenches, agriculture) suggest Robert Smithson with the resources of the U.S. Army Corps of Engineers. Robert Wood, a physicist who made major contributions to both ultraviolet light and ultrasound research, proposed a system of black cloth baffles, miles on a side, built in the desert and opened and closed by motor—a grid of pixels that could send "a series of winks" to observing Martians (*Popular Science* 1919, 75). Camille Flammarion argued for vast tracts of electric lights built in the Sahara, shining upward when Mars was in opposition. An A. Mercier, a colleague of Flammarion's, proposed constructing an enormous mirror or electric light in the heart of Paris (on the Champ-de-Mars, no less) (Frollo 1899). Alternatively—assuming that there might be some conservative opposition to constructing the brightest light on the planet, at the scale of the Brooklyn Bridge, in the heart of a densely inhabited metropolis—he suggested installing two mirrors on a mountain so that sunset light would reflect onto the shadowed side oriented to flash up at Mars, taking advantage of the dark background to use the light for better effect (Crowe 1986, 397).

Over the course of decades, two recurring landscape-as-medium proposals, generally but apocryphally attributed to the astronomer Joseph von Littrow and the mathematician Carl Friedrich Gauss, kept reappearing: to dig canals in the Sahara (circles or squares), fill them with kerosene, and set them on fire at night or to plant agricultural tracts in Siberia laid out as the square of the hypotenuse (the "windmill" diagram) (Crowe 1986, 205). Konstantin Tsiolkovsky, the great pioneer of rocketry (and advocate for the cosmic future of the human species), envisioned yet more tracts of mirrors (Crowe 1986, 397). Had there been a bit more free capital around for the turn-of-the-century "Mars mania," we might have some vast field of dusty, angled mirrors abandoned and reflecting the empty sky on a desert plateau somewhere, a Ballardian monument to the void.

All of these were more or less *visual* media projects, expressions of an era devoted to the idea of a crowded and lively solar system with canals on Mars, cloud forests on Venus, and underground populations of selenites on the moon. Franz von Paula Gruithuisen thought the famous "ashen light" sometimes observed on dark Venus was the product of "general festivals of fire given by the Venusians," with the forty-seven years between observations reflecting "the reign of an absolute monarch" before "another Alexander or Napoleon comes to supreme power on Venus." (Gruithuisen is sadly but accurately described by Crowe [1986, 204] as "a man of vast energy, extensive learning, excellent eyesight and instrumentation, and little sense.") Perhaps, speculated Mercier, the "flashes" observed on Mars were a response to the dazzling lighting of the Universal Exposition of Paris in 1889 (Crowe 1986, 397). When A. E. Douglass noted a "projection" on the Martian surface, which was reported in the press as another attempt at communication with Earth, he received theories as to the architecture of the population that could create such an effect: "Suppose the people of Mars have built a monument 10 miles square and a hundred miles high, covered exteriorly with polished marble" (Crowe 1986, 398).

If the populated universe were so neighborly, surely we could build tools by analogy to the Chappe semaphore telegraph, using basic visual symbols to convey the fact of existence and perhaps more complex information. Contacting the moon or Mars this way would be of a piece with broader technological transformations of the media apparatus, wrote Flammarion: "it is, perhaps, less bold than that of the telephone, or the phonograph, or the photophone, or the kinetograph" (Crowe 1986, 395). Even Gauss, much less prone to flights of fancy like his colleague Gruithuisen's "mad chatter," considered the use of the heliotrope—the mirror apparatus for reflecting sunlight over long distances to mark positions for surveying—for signaling: "This would be a discovery even greater than that of America, if we could get in touch with our neighbors on the moon." (In March 1822, with some back-of-the-envelope estimation, Gauss envisioned a hundred sixteen-square-foot mirrors used together—which would have been a splendid object, a fit companion to the Jantar Mantar architectural–astronomical buildings in Jaipur [Crowe 1986, 207].)

The visual component of these early projects tended to duck the follow-on question of what, precisely, was to be signaled and how it would be conveyed. All we could do was provide some evidence of our existence—a flashing light—and then presumably the acknowl-edgment would set the terms of the conversation. When the issue of the identity of our interlocutors was raised at all, it was under-stood that they would be more or less like us: after all, did they not dig canals and irrigation trenches? For purposes of communication, we could assume they were further along our inevitable historical and technological trajectory, "far superior to us" as Flammarion put it (Crowe 1986, 395). "Perhaps we will learn from an older and wiser planet how we ought to run the Earth," as *Popular Science* (1919, 74) had it with reference to telling the Martians about the just-concluded First World War. Only a few would-be stellar communicants consider the possibility of an alien biology. Francis Galton (1896, 661) has a hypothetical Martian transmission cracked by a little girl who points out that it's in base-8, not base-10, be-cause Martians count with six limbs and two antennae, like "highly developed ants," rather than ten fingers and toes. Gauss, rigorously imaginative as ever, "considering the universal nature of matter," hypothesized life on the sun with its massively higher gravity to consist of "only very tiny creatures . . . whereas our bodies would

be crushed" (Crowe 1986, 208). Guy Davenport once described
how seventeenth-century English translations of ancient Greek
make Achilles into a contemporary gentleman, always on the verge
of taking snuff; likewise, despite these steps away from anthro-
pomorphism, the assumptions of those plotting miles of mirrors
on mountain slopes are that Martians will be, more or less, as we
are—that we are addressing advanced Kants whose study windows
happen to look out on canals at the foot of Olympus Mons. The
conversation will be "begun by means of such mathematical con-
templations and ideas, as we and they have in common," suggested
Gauss (Crowe 1986, 206).

Even here, though, a more specific and abiding problem is becoming apparent: How are we to communicate anything more complex
than a flash of light or a right triangle? What does "communication"
mean, in this instance? Once we build those grids of electric lights
in the high desert, we can't just use "the Morse Code for it were
idle to suppose the Martians are familiar with this" (Crowe 1986,
400). In a problem with echoes of the Turing test or the Imitation
Game, cruelly inverted, how do we distinguish human activities and
attempts to communicate from the effects of natural events for an
unknown observer—volcanoes, auroras, bioluminescent seas, and
the radio hubbub of the universe at large?

That remark about Morse code is pertinent, because it is at this
point that the project of alien communication begins to shift from
the kinds of signaling techniques familiar to lost campers and
marooned sailors—flash a mirror, light a signal fire, arrange rocks
in a geometric shape—to richer, more abstract problems of communication that speak to the challenges of digital systems. As with
schizophrenic–paranoid anxieties, our models for human–alien
communication closely track developments in media technology.
Just after the turn of the century, Tesla (1901, 5) reported picking up signals from Mars (or possibly Venus) while building his
experimental wireless transmission apparatus in Colorado Springs,
with "a clear suggestion of number and order." He announced
the dawn of this new age in an article, "Talking with the Planets,"

which included a few lines that perfectly capture the combination of cosmic grandeur and can-do pragmatism that would come to characterize many of these communication projects: "with an expenditure not exceeding two thousand horsepower, signals can be transmitted to a planet such as Mars with as much exactness and certitude as we now send messages by wire from New York to Philadelphia" (4). The astronomer David Peck Todd, in 1909, proposed lifting "the most sensitive wireless telegraph receiver available" into the upper atmosphere with a balloon to pick up extraplanetary signals (Crowe 1986, 399). Through all of this, the problem of *what* to communicate—indeed, how to appear as a communicative act—remained uncertain. Tesla (1901, 5) breezily confirmed that communication, initiated by "a mere interchange of numbers," would rapidly move to "more intelligible" forms.

Part of the answer begins with the challenges presented by communicating using the new technical media themselves. There was a kind of symmetry between the two different forms of nonhuman communication.

Cros's Étude: Designs as Number Series

Charles Cros is now known—if he is known at all—as one of media history's also-rans, or as the author of some delightfully frustrating poetry. He invented techniques for three-color photography and a version of the phonograph, both of which he registered more or less immediately contemporary to other, more successful projects. (His phonograph—called, beautifully, the paleophone, *voix de passé,* the "past's voice"—was in many ways an Edisonian foil phonograph, not quite at the prototype stage, just as Edison was rolling out his first model.) He was one of the circle of *hydropathes,* artists and writers who shared a *fumiste* attitude combining outrageous subject matter with deadpan style, the result of a mocking, cryptic, too-cool sensibility that throws up a smokescreen and baffles outsiders—much of Erik Satie's dry amusement, his piano pieces played "like a nightingale with a toothache," is very *fumiste*—and specialized in grating nonsense poetry, like "The Salt Herring," to be read aloud to restive audiences. (The very model of a nineteenth-century inventor-dilettante, he also spent time on a project to manufacture fake jewels [Cros 1970, 541].) When not engaging in acts of deliberately failed communication with other humans, Cros was petitioning the French government to construct a huge and technically infeasible Archimedean burning mirror to etch shapes onto the deserts of Mars as a communications initiative. Concealed within the work of a man seemingly custom built for a cabinet of historical curiosities, however, was something far more profound.

In "Étude sur les moyens de communication avec les planètes," Cros (1970, 519) begins to seriously consider the challenge of reciprocal communication with an alien intelligence. He takes up the basic concept now well established—an enormous mirror flashing light to be seen by an observer on another planet—but asks how information is to be conveyed once the lines of communication have been opened. He considers, first, how a sequence of rhythmic flashes could be used to encode numbers but then takes up the question of whether those numbers could in turn encode *images.* A series of digits could communicate binary pixels—spaces black or white, off or on—in lines on an ordered grid, in the style of "6–1 2–0 3–1 7–0" for

XXXXXX00XXX0000000

using integers rather than having to flash all those signals one by one. (Cros devotes some time to how exactly this message-sending protocol would be initially communicated.) As he outlines his project, it becomes clear to the modern reader that he has developed a version of what is now called "run-length encoding," an image compression and transmission technology akin to that used in fax machines, early digital bitmap images, and some of the very first television technologies collected by Siegfried Zielinski. There would need to be encoding systems for turning images—and, potentially, other kinds of media—into materials for this notation-transmission apparatus: "analogous notation procedures for rendering designs as number series are used in various industries, including weaving and embroidery."

At this point, the ears of historians of computing might prick up: what kind of industrial weaving machines, pray tell? "There is, in [Jacquard weaving], a whole science that, as so often happens, was practiced before it was theorized. From it will emerge a new and important branch of mathematics, and eventually a new classification of these primordial sciences [i.e., the sciences of information and data storage]. The study of rhythms [patterns and encoding systems] will take its place alongside that of figures" (Cros 1970, 534). In context, Cros's "study of rhythms" means a set of instructions to be carried out in a particular order on the machine's material: what we would now call an algorithm. What we have here is a project to turn Earth into a graphics card, encoding images and eventually other data for transmission to be rendered and displayed elsewhere. The project of developing nonhuman communication, with its problems of abstraction, encoding, compression, error correction, and display, turns out to be analogous to the problem of developing computable media—to "what we now call programming."

Astraglossa: How to Point at Things

"What we now call": it was new then, in 1952, when the zoologist and medical statistician Lancelot Hogben wrote "Astraglossa," a lighthearted but extremely thorough and detailed study of the format of potential extraterrestrial communication. During the Second World War, Hogben (1943) had published "Interglossa," a proposal for an auxiliary language with an inventively simple structure—a kind of international argot of science and technology inspired by the complexities of teaching biology to a cosmopolitan student body. Astraglossa is something else entirely: not a language as such but an analysis of what it means to communicate with a nonhuman, unknowable interlocutor. Prior to meaning as such, prior to language, Hogben—who in his working life was occupied with hormonal signals of African clawed frogs and color-

shifting reptiles and amphibians—was interested in the most **33** minimal order of signaling: "a technique of how to point at things" (Hogben 1963, 126). If we assume (and this is already a large and complex assumption) that those we seek to contact share a sense of *time*—and with it number, interval, and the stars—then we can produce what for Hogben constitutes the fundamental structure of this most minimal nonhuman communication: rank order, gaps, and iteration. That is, if all you can communicate are sequences of electromagnetic pulses, dots and dashes, you rely on order in *time*. A shorter time interval separates a chunk of pulses meant to be taken conceptually together, and a longer interval marks the conclusion of a linked series of such chunks—*gaps*. Their sequence in time establishes the role of different sequences of pulses—*rank order*. One follows another, with the same or different operations executed repeatedly to arrive at a result—*iteration*:

    1 .. Fa .. 1.1 .. Fa .. 1.1.1 .. Fb .. 1.1.1.1.1

or one plus two plus three equals six, with the periods standing for units of time between pulses. The *F*s in Hogben's notation refer to "flashes," sequences of pulses with distinctive properties whose placement by gaps and rank order suggests the operation of addition and identification or equality.

We have pulses and the time between pulses, some of which we can arrange into operations—addition, subtraction, identity, affirmation and negation, and so on—and we can then, given enough time, stack the operations into "flashes" that constitute *rules*: apply the set of operations collected by this rule to the following string of numbers. Followed by silence. Which thus becomes the framework for a signal, a flash, corresponding to elicitation or a question— awaiting the product of an operation performed. All of this should begin to sound rather familiar: "The only unmentioned clue that I regard as specially relevant to our theme I shall merely refer to *en passant,* viz. what we now call programming, i.e. the syntax of the language in which we transmit orders to the new electronic computing machines."

The parallels that Hogben (1963) draws are illustrative not only of his thinking about this subject but of the larger challenges of finding the edges of the concept of communication—the places where we enter and leave the communicable, what he calls the "common field of semantic reference" (124). He draws on the history of interpreting the Mayan glyphs—which, in 1952 as he composed these ideas, had only encompassed the numerical and astronomical and calendrical systems present in the language. He compares the project to the universal visual languages of Otto Neurath's ISOTYPE and Charles Bliss's semantography (or as it is now known, Blissymboblics)—both of which build symbolic vocabularies that don't correspond to sounds but instead express concepts wholly through visual objects and abstract operators. He plays at length on the theme of childhood, infancy, and education: how do we try to teach shared symbol systems to children? He compares the puzzles of establishing concepts like antitheses, interrogatives, and assents to "a fuller case history of Helen Keller" (131). All of these analogies are laden and complex, and it would be very interesting to consider the different models of "communication" they imply—but most interesting for our purposes is the challenge of alien communication expressed as an essentially *computational* problem.

All of his other comparisons rely on human commonalities, from the shared sky over Mayan Mesoamerica to the silhouettes and color codes of ISOTYPE to Keller's understanding and Anne Sullivan's hands, making this computational aspect stand out all the more. The challenge of establishing a shared binary symbol system and logic of operations, predicated on gaps, rank order, and iteration, closely resembles the work done by Turing, Tom Kilburn, Freddie Williams, and others to build Turing-complete electronic computers in Manchester and London contemporaneously with Hogben's talk. (Manchester happens to be the city where Hogben sets his fictional classroom of analogical Martian pupils.) Finding the minimal fundamentals necessary to express more complex ideas has parallels with Boolean algebra and logical processes like

Charles Sanders Peirce's (1989) demonstration that NOR gates, properly arranged, can produce the functions of all logic gates and therefore of a functionally complete logical system (what came to be called "Peirce's arrow").

Finally and crucially, however, this is not the full extent of Hogben's ambition. He does not want to establish logic for its own sake, to produce "a monologue of simple assertions," but as a step toward a rapport. The later parts of his study playfully but carefully analyze how he could establish pronouns (*your* and *our, it* and *they, I*-ness); terms of assent, denial, and doubt; conditionals and assertions; causes and consequences, entirely within a system of binary pulses and "flash" operators with reference to time and stellar objects. The ultimate goal is to use the framework of this logic for "reciprocal communication" with the unknown, as "our Neolithic forebears . . . can communicate with us" through numerical and calendrical relics like notched bones and standing stones, or as we "transmit orders to the new electronic computing machines."

Of course, even this elegant attempt at a minimal set of communication components relies on some necessary assumptions as to how it will be interpreted and understood.

"How does it come about that this arrow >>>--> *points*? Doesn't it seem to carry in it something besides itself?" So asks Wittgenstein (2001), in section 454 of the *Philosophical Investigations*. Why, when someone points, do we assume that the direction runs from elbow to fingertip and not the other way around? (Or, as Laurie Anderson [1983] asked of the waving man depicted on the Pioneer Plaque, the engraved plates of aluminum mounted on the Pioneer spacecraft in the hope of being intercepted by extraterrestrials: "Do you think that they will think his arm is permanently attached in this position?") Wittgenstein notes, "Perhaps a Martian would describe the picture so." How would our unknown audience *listen* to something, for instance? Again, the challenge of formatting extraterrestrial conversation takes us directly to the puzzles of communication in technological media.

Set aside, for a moment, the complexities entailed by establishing
the primitives of formal logic using only radio pulses beamed
to some corner of the sky with decades-long time lag. Instead
just imagine a record—an analog LP. Take it for granted that the
recipients will be able to decipher the instructions for building the
player or will invent their own. What do we put on the record to
express our experience?

Voyager and Arecibo: A Martian Would
Describe the Picture So

That's the most common question for such a project: we jump to
the content of the communication—the usual Bach-or-Fela Kuti-or-
a-child's-laugh debate about which object could stand in for many
objects, which unit expresses a category. More difficult and much
more profound, however, is the question of *format*: what about this
groove etched in an anodized disk conveys that this is "our" "experience" prior to any particular chunk of that experience? Just as
conducting even the most minimal conversation about astronomical
states requires us to reinvent time and spatial notation, *we-you* pronouns, direction, and ordinality, so the commonsensicality of sound
demands deep consideration, one which echoes Kittler's work.

Imagine two records, then. One is made of gold-plated copper
from 1977, currently in interstellar space outside the heliosheath.
The other is a dictation cylinder produced in 1900. Kittler has
written about the latter, emphasizing one of the most significant
and least obvious things about it: that it records the *voice,* not
just "the words." It is a mechanical process, transforming what is
said not into words, as a scribe would, but into data storage—the
stylus records the voice aspeak but also the birds singing outside,
the creaking of the floorboards underfoot, and of course the
mechanical noise of the recording medium itself. It shifts the
boundaries of what can be inscribed and recorded—what becomes
part of the record, part of discourse. Along with other novel media
technologies, it creates a new form of human expression that

sounds familiar to our present data-driven order: that we expose what is most personal and individual to us not through what we deliberately *say* but through what the machines pick up about us. We are our fingerprints, unconscious microexpressions captured on film, our tone of voice and mistakes of speech and background sounds. (From *Blow-Up* to *Blow Out* to *The Conversation* to *The Girl with the Dragon Tattoo,* a new genre has come into being based on the tales told by machines recording events whose significance at first escapes their human users.) "Only the phonograph can record all the noise produced by the larynx prior to any semiotic order and linguistic meaning" (Kittler 1999, 16). And only for humans would all that extraneous noise be so *invisible*: the gray velvet mounts on which the gleaming jewelry of spoken language is displayed, when we encounter it through the mediation of tinfoil and shellac, nails and needles and resonant cones.

Jon Lomberg, reflecting on the production of the Voyager "Golden Disk" in 1977, articulated the problem (Lemarchand and Lomberg 2011). The record was to include music, human speech, technological sounds, animal sounds, and natural sounds produced by the planet, addressed to unknown alien listeners. Within the format of the recording itself, Lomberg and his colleagues wanted to convey a set of profound and subtle distinctions. First, there were distinctions in "nature of the sound" between the music and the other sounds. The music was to be *experienced* and analyzed for its structure, but the other sounds were to be *identified* for their informational content. Music expresses subjective states; the other sounds are objective conditions. (I ask the reader's indulgence in these deep waters about the questions already raised—that human sounds, music included, are not a product of nature—just to consider the format and communication problems.) Furthermore, and still more delicately: how to communicate that the speech (greetings in different languages, children talking, party noise), technology (hammering nails, engines turning over), and music are all "our" sounds and birdsong, bees buzzing, a roll of thunder, dogs barking, and whalesong are *not*?

Lomberg sought to connect human activities with the sound of a heartbeat, running it as a backing track to distinguish "our" sounds from all the others—and to produce a kind of sequence, a variant of Hogben's use of rank order and gaps, with the technological sounds as a series of acoustic narratives that would build from running feet through internal combustion to a jet engine and a rocket launch. Indeed, Lomberg exploited the very property of audio recording that Kittler emphasizes: that the voice as inscribed is no longer a matter of words understood, interpreted, and written but a bodily event of moderated breath in the larynx. The recordings of speech should be made "so that an intake of breath before syllables could be heard. This would link breathing with speech, and perhaps give a clue as to the respiratory nature of speech, and link the sounds of speech with the heartbeat" (Lemarchand and Lomberg 2011, 379). The heartbeat was a kind of metadata about a particular set of sounds. Given the fearsome complexities and challenges in interstellar linguistics, perhaps the most important and salient component of our recorded speech is precisely that it is anatomical first and foremost.

Every extraterrestrial communication project forces *two* assertions: the minimal requirements for communication and the most significant matter to be communicated. With a few exceptions, like the Voyager record, the minimal requirements are *pulses,* binary strings of energy encoding different kinds of messages or representational schemes—in other words, systems more or less directly inspired by or in dialogue with computing and telecommunications projects. Cros was a forerunner, well ahead of the state of the art in adapting notation from Jacquard looms to produce a celestial fax machine; by the 1970s, with the famous Arecibo message, Frank Drake (1992) had worked out a string of 1,679 bits (on and off pulses, repeating) to be sent from the radiotelescope of the same name in Puerto Rico. The number 1,679 is semiprime, the product of the primes 23 and 73, and if you arrange the on and off signals in order in a grid of twenty-three columns and seventy-three rows, you have a picture—with the pulses as the pixels.

And what content? What is the substance of communication
with unknown and unknowable interlocutors to be? A minimal
set of facts, almost always: a numbering system, a set of stellar
coordinates, a few facts of chemistry, a human silhouette. Arecibo,
read top to bottom, provided numbers in binary, the atomic
weight of the basic elements in our biology and the chemistry of
DNA, our population and physical shape (the wavelength of the
message itself provides the scale), the arrangement of our solar
system, and finally the antenna itself. Most such messages are
necessarily humble, primarily concerned with the structure of their
own decoding—after all, the simple, phatic fact of a *we are here*
statement is itself of enormous consequence. Successful use of the
medium, with a signal distinguishable from the electromagnetic
activity of the universe in general, is the event. The content can be
more or less of the "Watson—come here—I want you" / "What hath
God wrought?" / "This has been a day of solid achievement" variety
(for telephone, telegraph, and hard drive, respectively).

Even here though, beaming minimal signals to distant stars and
"missing subjects," we find ambitious projects to expand on what
the format is capable of—and we reconnect completely with the
question of rethinking "communication" in terms of our own
nonhuman and alien media technologies. What can we communicate using this set of binary pulses and logical primitives? What
are the limits? People have built everything from more complex
Arecibo-like messages, with new pixel arrangements (Zaitsev and
Ignatov 1999). The "CosmicOS," a "self-contained message" made
of four symbols that account for binary code and brackets, acts
as a kind of computer program to be executed. (It looks like this:
◇|◇◇◇◇||◇⌊□⌉ . . .) This last example, while also playful, is exact
and thorough and speaks directly to the challenge of thinking past
our own biology in our communication efforts. One of the goals of
CosmicOS is

> to avoid making too many assumptions about the percep-
> tual abilities of the non-human intelligence; for example
> that they make sense of 2D images in the same way we

do. While some arguments can be made for this, as a machine vision researcher I am very skeptical that we really understand the variability possible here. (Fitzpatrick, n.d.)

It is through thinking about how machines can "see"—or can use novel techniques to do something like seeing—that we can also understand how different forms of life might see, or sense, or otherwise interact with visible light.

CosmicOS builds on another, earlier experiment in pushing the limits of possible extraterrestrial communication: the constructed language Lincos. It is one of the most rigorously eccentric intellectual projects of the twentieth century, to put "in principle the whole bulk of our knowledge" into a form communicable to any possible intelligent life, and we will close this chapter with it.

Lingua Cosmica: Human, or at Least Humanlike

"He begins with elementary mathematics," writes Marvin Minksy (1985) of the Lincos project, "and shows how many other ideas, including social ideas, might be based on that foundation." Minsky, one of the major figures in AI research—cofounder of MIT's AI laboratory, author of *The Society of Mind,* advisor to Clarke and Kubrick in the production of *2001*—is describing the work of the mathematician Hans Freudenthal. The lingua cosmica, or Lincos, is a language that begins with "peeps" of discrete radio pulses for the natural numbers and ends with relativistic mechanics. On the way, it includes set theory, cardinality and ordinality, assertions ("Future events *cannot* be perceived"), a "short history of Fermat's theorem," "examples of polite speech," bets and gambling, the act of wishing, points and vectors, and "whistling for one's dog." ("The dog refuses.") And this was to be only part one of a projected two, with the second including chapters on "Matter," "Earth," and "Life" and an additional supplement on "Behavior" to build on the first book (Freudenthal 1960, 23). All of these contributions are rendered in an increasingly complex formal notation, express-

ing how they would be transmitted as radio pulses. It looks like
this:

   ↔PauAnt•*He*Dat*Hd*.Den0,101▐

Once we are past the same basics touched on in one way or
another by every broadcast concept more complex than Arecibo—
numbers, spatial coordinates, timing, basic logical and mathematical operations—Freudenthal, like Hogben, wants to do something
far more ambitious than simple declarations about the nature
of things. He establishes a set of human actors and embarks on
a series of logical–minimalist playlets. These conversations and
events between *Ha* and *Hb*—described entirely in Freudenthal's
formal notation—establish stories about the nature of the world
and, more to the point, about the nature of human experience in
its most austere form. *Ha* throws a ball farther than *Hb* can catch it.
Hb knows something but does not say it, which means that *Ha* does
not know it; *Ha* can try to guess what it is that *Hb* knows. *Ha* and *Hb*
know what happened in the past but not what will happen in the
future, and they bet on an outcome together. *Ha* didn't see something and therefore asks *Hb* about it. Together, they live in a world,
and there are many other things that live in the world with them
with whom they can't communicate in the same way, even though
those things can also see, hear, move, know the past, and chase a
ball. They can die, *Ha* and *Hb*; so can all the other things with which
they share the world. They can wish that things were otherwise
than they are. When one of them dies, they can no longer talk.

It in no way diminishes Freudenthal's strange achievement—an
attempt at formalizing human life in the universe into a basic set
of electromagnetic signals—to question its fitness for purpose: the
nearest possible life is so far away that years or decades would
separate each exchange confirming receipt and understanding
with reciprocal signals. Some of Freudenthal's dialogical units are
hundreds of steps long, with multiple points needing confirmation;
simply at the level of back and forth, this would be a millennial
project. What he has produced in fact seems much closer—much

more appropriate—to formulating properties of human experience not to communicate with aliens but with machines.

George Boole, walking across a field at seventeen in 1832, was struck by an idea that led to his book *An Investigation of the Laws of Thought,* a model of logical reasoning using a symbolic calculus built on what he called "the Universe" and "Nothing"—or 1 and 0 (Boole 2009). What he produced was not an accurate portrait of the process of human reasoning, but it was, a century or so later, perfectly suited to performing complex mathematical and logical operations on electrical relays and vacuum tubes. Likewise, Freudenthal's project is filled with striking insights, and many of them seem less applicable to broadcasting to Alpha Centauri than to explaining what it is to be a person in the universe to an entity that lacks anything but a memory and the input of a very limited set of electromagnetic symbols. No wonder Marvin Minsky, who constructed the first neural network simulator, was drawn to this research: it shoots for the stars and lands in the AI lab.

Indeed, one of the persistent challenges of AI has been precisely in recognizing the particular biases and tendencies in human cognition that would not be shared by a machine—no more than the way we think that we see is shared or modeled by a sensor picking up photons. "I shall suppose that the person who is to receive my messages is human or at least humanlike as to his mental state and experiences," writes Freudenthal (1960, 14). "I should not know how to communicate with an individual who does not fulfill these requirements." Now we do so all the time: by voice, text, image, and the indirect production of data.

We have established rapport with an alien planet that we build and maintain around us, teaching its population to make spatial sense of the world, keep secrets, recognize faces, hear and compress and filter voices, make conversation, and interpret a far broader range of electromagnetic radiation than just radio waves or the visual spectrum. We have done all this with binary pulses, logical operations, and encoding and decoding schemes.

The informational commerce and exchange of the world take place in the dense network of Martian canals that we've dug for the last sixty years, and we communicate in ways that embody the problems and solutions of Cros and Gauss, Hogben and Lomberg and Freudenthal. Like fantastical narratives that act as accounts of our society alienated and rendered strange—from *Gulliver's Travels* to *Animal Farm*—the three centuries of projects for communicating with unknowable, nonhuman interlocutors chronicled here can be rearranged (*Erewhon* becomes Nowhere) to be our most ordinary present: reacting to automated alerts, talking with customer service, solving CAPTCHAs to log in to Facebook.

This is our present state of mediated communication. What is our future?

The Light Cone

Hogben was not alone in this conceptual move, but he put it most eloquently: "Though we cannot communicate with our Neolithic forbears, they themselves can communicate with us through the mists of time"—with "astrocalendrical" hieroglyphs and number systems. Many extraterrestrial communication projects make reference to the case of our distant ancestors and their notched bones and standing stones marking, or appearing to mark, lunar sequence and equinoctial sunrise. With a basic vocabulary of forms, used to track regular cosmic patterns, they were able to establish a one-way communication channel over thousands of years. It seems similar, in some ways, to a message broadcast out into the void. But this position is riddled with problems if it is taken any more seriously than a straightforward statement on the value of numbers and astronomical events as points of common reference. After all, we enjoy anatomy, fundamental cognitive and social traits, and a planet in common. Why do we use ourselves in the past as a way of analogizing communication with alien life in the future?

This chapter has been concerned with the idea that, in considering communication in digital media, we should account for how alien, and how ubiquitous and invisible, the interlocutors and mediators we built have become. This corresponds to a dehumanization—again, not in a pejorative sense!—of contemporary digital media. I find the analogy of extraterrestrial communication useful in seeing this problem, because it forces questions about the anthropocentric sense of communication on digital networks and acts as a kind of parallel history of digital networks themselves and the formats we've developed to make use of them. Toward the end of the first section, I asserted that we could move beyond the models of Kittler and Licklider and Taylor to look at communication with the future as well as past and present—a future defined by communications media that are not just digital but *alien.* Rather than hiding their profound strangeness behind carefully designed interfaces and media that, to quote Bunz, "address us as if we were children," complete with cartoon characters, bright flat design, and tinkly-tonkly toy instrument sound tracks, we should expose and explore—and even embrace—their alienness and, with it, the future.

We have talked about theories defining the limits of information, like Claude Shannon's, but there is one more limit to communication: a temporal one. The light cone is the theoretical description of an event—a flash of light, a signal—propagating through spacetime. What lies within the cone could have been influenced by the signal in some way. What lies within the *future* light cone, and could therefore receive a signal from a present moment, is the set of all future events that could be causally influenced by it. It's a steadily expanding zone of possible causality and a way of understanding the temporality of information and communication. You can understand an event taking place here and now as being within the light cone of many remote objects in the universe, to the edge of observation itself—but as they were, of course, at the moment of an equally remote event, and not now. We are in their (far, far) future.

"To get information about the life and times for some inhabitants of a planet outside the Solar system, say, the fastest way that we can get such information is by means of light signals," write Grøn and Hervik (2007). "The light-cone tells us what region of spacetime we can get information from" (221). Both the way light moves and the speed at which it moves are hard limits on what can be known. All the emissions of attempted communication with alien life—the mostly hypothetical flashes of light, the bursts of radio signals—are light cone events, utterances in space-time. That is, they are communications not simply with a distant exoplanet but with our future.

If they are intercepted at all, these messages will arrive years—or more likely decades, millennia, or eons—later. Given time for interpretation and response, that almost unimaginable conversation will likely take place after the original human speakers are dead. We likewise address ourselves to our present, in the act of communication, but also to a future likely populated with far more, and more diverse, varieties of alien interlocutors—the "missing subjects" who await our signals and whose predecessors we are programming, building, and installing on Earth at the moment.

I would like to end on this note of depersonalized optimism: that our present communicative acts on and with digital infrastructure put us in conversation not only with one another, or with our current machinery, but also with far stranger communicants that will have as little resemblance to our media experience as Spanner does to the CTSS. Future technical infrastructures are what we digitally communicate with now, infrastructures far more alien in their likely operation, and in the place they occupy in the future itself. This is how we can retrospectively understand ourselves and our communication using contemporary digital media: messages formatted and transmitted to our alien planet to come, a little way down the light cone, close enough to guess at but unimaginably far away.

Note

I owe many thanks for this project, starting with Wendy Chun's invitation to join the Terms of Media II conference. The question and discussion session during that event was invaluable. This piece was informed throughout by the work and conversation of Mercedes Bunz and Sara Dean. My thanks in particular to Boris Traue for a thoughtful, informed, and thorough editorial reading of the manuscript, which improved it significantly; to Paula Bialski for analysis, suggestions, and the introduction; and to the staff of meson press and the insights of the peer reviewers. My fond gratitude to you all.

References

Ades, Dawn, ed. 2006. *The DADA Reader: A Critical Anthology.* Chicago: University of Chicago Press.

Anderson, Laurie. 1983. "Say Hello." *United States Live.* New York: Warner Brothers Records.

Bardini, Thierry. 2000. *Bootstrapping: Douglas Engelbart, Coevolution, and the Origins of Personal Computing.* Stanford, Calif.: Stanford University Press.

Bartlett, Jamie. 2017. "Can You Distinguish between a Bot and a Human?" *Spectator,* November 20.

Boole, George. 2009. *An Investigation of the Laws of Thought on Which Are Founded the Mathematical Theories of Logic and Probabilities.* Cambridge: Cambridge University Press.

Bratton, Benjamin. 2016. *The Stack: On Software and Sovereignty.* Cambridge, Mass.: MIT Press.

Brunton, Finn. 2013. *Spam: A Shadow History of the Internet.* Cambridge, Mass.: MIT Press.

Canales, Jimena. 2009. *A Tenth of a Second: A History.* Chicago: University of Chicago Press.

Capova, Klara. 2013. "The Charming Science of the Other: The Ethnography of the Scientific Search for Life beyond Earth." PhD diss., Durham University.

Chun, Wendy. 2008. "The Enduring Ephemeral, or the Future Is a Memory." *Critical Inquiry* 35: 148–71.

Corbett, James C., Jeffrey Dean, Michael Epstein, Andrew Fikes, Christopher Frost, J. J. Furman, Sanjay Ghemawat et al. 2012. "Spanner: Google's Globally-Distributed Database." In *Proceedings of OSDI 2012.* Accessed March 25, 2016. http://static .googleusercontent.com/media/research.google.com/en//archive/spanner -osdi2012.pdf.

Cros, Charles. 1970. *Oeuvres complètes.* Paris: Pléiade.

Crowe, Michael J. 1986. *The Extraterrestrial Life Debate 1750–1900: The Idea of a Plurality of Worlds from Kant to Lowell.* Cambridge: Cambridge University Press.

Drake, Frank, and Dava Sobel. 1992. *Is Anyone Out There?* New York: Delacorte Press.

Edwards, Paul. 2003. "Infrastructure and Modernity: Force, Time, and Social Organization in the History of Sociotechnical Systems." In *Modernity and Technology,* edited by T. Misa, P. Bray, and A. Feenberg, 185–225. Cambridge, Mass.: MIT Press.

Engelbart, Doug. 1962. *Augmenting Human Intellect: A Conceptual Framework.* Report to the Director of Information Sciences, Air Force Office of Scientific Research. Menlo Park, Calif.: Stanford Research Institute.

Eno, Brian. 1996. *A Year with Swollen Appendices.* London: Faber and Faber.

Fielding, Nick, and Ian Cobain. 2011. "Revealed: US Spy Operation That Manipulates Social Media." *Guardian,* March 17.

Fitzpatrick, Paul. N.d. "CosmicOS." CSAIL. Accessed March 4, 2016. http://people.csail .mit.edu/paulfitz/cosmicos.shtml.

Foucault, Michel. 1990. *The History of Sexuality: Vol. 1. An Introduction.* New York: Vintage.

Freudenthal, Hans. 1960. *Lincos: Design of a Language for Cosmic Intercourse.* Amsterdam: North-Holland.

Frollo, Jean. 1899. "De la Terre a Mars." *Le Petit Parisien,* May 10.

Galloway, Alexander. 2004. *Protocol: How Control Exists after Decentralization.* Cambridge, Mass.: MIT Press.

Galton, Francis. 1896. "Intelligible Signals between Neighboring Stars." *Fortnightly Review* 60: 657–64.

Goodhart, Charles. 1981. "Problems of Monetary Management: The U.K. Experience." In *Inflation, Depression, and Economic Policy in the West,* edited by Anthony S. Courakis, 91–121. Lanham, Md.: Rowman and Littlefield.

Gray, Robert H., and Simon Ellingsen. 2002. "A Search for Periodic Emissions at the Wow Locale." *Astrophysical Journal* 578, no. 2: 967–71.

Greenberger, Martin. 1962. *Management and the Computer of the Future.* Cambridge, Mass.: MIT Press.

Greenwald, Glenn. 2014. "How Covert Agents Infiltrate the Internet to Manipulate, Deceive, and Destroy Reputations." *Intercept,* February 24.

Gregg, Melissa. 2013. "Courting Vectoralists: An Interview with McKenzie Wark on the 10 Year Anniversary of 'A Hacker Manifesto.'" *Los Angeles Review of Books,* December 17.

Grøn, Øyvind, and Sigbjorn Hervik. 2007. *Einstein's General Theory of Relativity: With Modern Applications in Cosmology.* New York: Springer.

Hogben, Lancelot. 1943. *Interglossa: A Draft of an Auxiliary for a Democratic World Order, Being an Attempt to Apply Semantic Principles to Language Design.* New York: Penguin Books.

Hogben, Lancelot. 1963. *Science in Authority.* New York: W. W. Norton.

Internet Assigned Numbers Authority. 2016. "Time Zone Database." Accessed March 25, 2016. https://www.iana.org/time-zones.

Irani, Lilly. 2015. "Justice for 'Data Janitors.'" *Public Books,* January 15.

Jain, Anab. 2014. "Valley of the Meatpuppets." Superflux. Accessed March 25, 2016. http://superflux.in/index.php/work/valley-of-the-meatpuppets/#.

Kittler, Friedrich. 1990. *Discourse Networks 1800/1900.* Stanford, Calif.: Stanford University Press.

Kittler, Friedrich. 1999. *Gramophone, Film, Typewriter.* Stanford, Calif.: Stanford University Press.

Kittler, Friedrich. 2010. *Optical Media.* Cambridge: Polity Press.

Krebs, Brian. 2015. "Extortionists Target Ashley Madison Users." *Krebs on Security* (blog). August 15. https://krebsonsecurity.com/2015/08/extortionists-target-ashley -madison-users/.

48 Kuo, Michelle. 2008. "Special Effects: Michelle Kuo Speaks with Michael Callahan about USCO." *Artforum,* May 2008.

Lee, John A. N., and Robert Rosin. 1992. "The Project MAC Interviews." *IEEE Annals of the History of Computing* 14, no. 2: 14–35.

Lemarchand, Guillermo A., and Jon Lomberg. 2011. "Communication among Interstellar Intelligent Species: A Search for Universal Cognitive Maps." In *Communication with Extraterrestrial Intelligence (CETI),* edited by Douglas A. Vakoch, 371–95. Albany: State University of New York Press.

Licklider, J. C. R. 1963. "The System System." In *Human Factors in Technology,* edited by E. Bennett, J. Degan, and J. Spiegel, 627–45. New York: McGraw-Hill.

Licklider, J. C. R. 1965. *Libraries of the Future.* Cambridge, Mass.: MIT Press.

Licklider, J. C. R. 1988. "Oral History Interview by William Aspray and Arthur L. Norberg, 28 October 1988, Cambridge, Massachusetts (OH 150)." Charles Babbage Institute, University of Minnesota.

Licklider, John, and Robert Taylor. 1968. "The Computer as a Communication Device." *Science and Technology* 76: 21–31.

McLuhan, Marshall. 1994. *Understanding Media: The Extensions of Man.* Cambridge, Mass.: MIT Press.

Mills, D. L. 1985. "RFC 956: Algorithms for Synchronizing Network Clocks." Network Working Group. Accessed March 25, 2016. https://tools.ietf.org/html/rfc956.

Minsky, Marvin. 1985. "Communication with Alien Intelligence." In *Extraterrestrials: Science and Alien Intelligence,* edited by Edward Regis, 117–28. Cambridge: Cambridge University Press.

Negarestani, Reza. 2015. "Revolution Backwards: Functional Realization and Computational Implementation." In *Alleys of Your Mind: Augmented Intelligence and Its Traumas,* 139–54. Leuphana, Germany: meson press.

Newitz, Annalee. 2015. "How Ashley Madison Hid Its Fembot Con from Users and Investigators." *Gizmodo,* September 8.

Parikka, Jussi. 2015. *A Geology of Media.* Minneapolis: University of Minnesota Press.

Peirce, Charles Sanders. 1989. "A Boolean Algebra with One Constant." In *Writings of Charles S. Peirce,* vol. 4, 13–18. Indianapolis: Indiana University Press.

Peters, John Durham. 2015. *The Marvelous Clouds: Toward a Philosophy of Elemental Media.* Chicago: University of Chicago Press.

Popular Science Monthly. 1919. "Hello Mars—This Is the Earth! Will the Martians Answer Us?" September.

Sandvig, Christian. 2013. "The Internet as Infrastructure." In *The Oxford Handbook of Internet Studies,* edited by William Dutton, 86–106. Oxford: Oxford University Press.

Shannon, Claude. 1956. "The Bandwagon." *IRE Transactions on Information Theory* 2, no. 1: 3.

Shannon, Claude, and Warren Weaver. 1949. *The Mathematical Theory of Communication.* Chicago: University of Illinois Press.

Sindelar, Daisy. 2014. "The Kremlin's Troll Army." *Atlantic,* August 12.

Starosielski, Nicole. 2015. *The Undersea Network.* Durham, N.C.: Duke University Press.

Sterne, Jonathan. 2012. *MP3: The Meaning of a Format.* Durham, N.C.: Duke University Press.

Stites, Richard. 1989. *Revolutionary Dreams: Utopian Vision and Experimental Life in the*
   Russian Revolution. New York: Oxford University Press.

Tesla, Nikola. 1901. "Talking with the Planets." *Collier's Weekly,* February 9.

Turing, Alan. 1950. "Computing Machinery and Intelligence." *Mind* 59: 433–60.

Verraes, Mathias. 2015. "There Are Only Two Hard Problems . . ." Twitter. Accessed
   March 25, 2016. https://twitter.com/mathiasverraes/status/632260618599403520.

Vismann, Cornelia. 2008. *Files: Law and Media Technology.* Stanford, Calif.: Stanford
   University Press.

Waldrop, M. Mitchell. 2001. *The Dream Machine: J.C.R. Licklider and the Revolution That
   Made Computing Personal.* New York: Viking.

Wittgenstein, Ludwig. 2001. *Philosophical Investigations.* London: Blackwell.

Yates, JoAnne. 1989. *Control through Communication: The Rise of System in American
   Management.* Baltimore: Johns Hopkins University Press.

Zaitsev, Alexander, and Sergey Ignatov. 1999. "'Broadcast for Extra-Terrestrial Intelligence from Evpatoria Deep Space Center': Report on Cosmic Call 1999." Accessed
   March 19, 2016. http://www.cplire.ru/html/ra&sr/irm/report-1999.html.

2

The Force of Communication

Mercedes Bunz

The things around us, having become media, have started to
address us. Their first utterances went unnoticed: for years, our
cars have loudly insisted that we fasten our seat belts. Informed
by sensors, they scream as if they feared for their bodies while
being parked or shout for help when they reckon that someone
else, whom they do not know, wants to take them. This mode
of communication quickly spread to the house. Now the robotic
vacuum cleaner eagerly informs us when it is stuck and asks us to
"move Roomba to a new location." And driven by new advances
in natural language processing I have explored elsewhere (Bunz
and Meikle 2018, 45–67), intelligent personal assistants with
names like Siri and Alexa wake up to address us when they hear
someone calling their names—in contrast to our fellow humans,
who ignore everyone around them while under the spell of a
screen. When things became interactive, they established a new
kind of dialogue with us, the humans. To use technical interfaces
today means to communicate with technology. Of course, it is not
technology itself that has raised its head and started to speak.
Even though it has learned to communicate, it has not become a
human subject, although it has always been more than an object.
Heidegger ([1954] 1977, 4) had good reason to look *further* into
the agency of technology by reconsidering what is usually taken

for granted—"technology is a means to an end. . . . Technology is a human activity"—thereby questioning the instrumental definition of technology. Now that our technological devices have started to address us with multiple voices, we need to continue his analysis. So in what way can we investigate how technology addresses us without thinking it is speaking to us? For this is certain: when technology starts to speak, it is not technology we hear. Still, this is a development that is transforming our contemporary discourse and, with it, what can be called our "being with technology." This essay explores the force of digital communication, starting with a methodological discussion of how to approach technology. Having clarified this, it then links different aspects together: communication theories and the way we are addressed by digital media, child psychology and computer science, interface design and political theory. But let us start this endeavor by looking at what happens— what forces speak—when we communicate.

Being with Technology

Communication theories have always suspected that communicating with media *transforms our being in this world* in various ways. This section approaches these theories and this transformation in three ways. First, it summarizes historical theories of communication to foreground their common assumption, namely, that there is a *force* happening when we communicate. To understand where this force is generally located when it comes to digital technology, it then turns to contemporary theories. Finally, it discusses technology as a situation: the situation of being addressed by digital technology. But let's start with historic takes on communication.

Over the years, theorists have developed very different takes on communication. Yet, one assumption has always been at the heart of all theories: there is a force happening while we communicate. The following communication theories illustrate this, although the list is by no means exhaustive:

Shannon. An interest in the force of communication can already be noticed in one of the early theoretical takes on communication, in Claude Shannon and Warren Weaver's (1949) *The Mathematical Theory of Communication,* which my coauthor Finn Brunton discusses with brilliance and in more detail in chapter 1. Their theoretical concept of information implies that the capacity of a medium defines its possibilities to produce meaning, thereby claiming a certain dependency on the transmitting medium. Inspired by their theory, the German media theorist Friedrich Kittler (1999, xxxix) would condense this later to the claim that "media determine our situation, which—in spite or because of it—deserves a description."

Derrida. The French philosopher adds to this perspective (that something else is going on when communication is happening) by observing that communication also does not simply transmit content. As he points out in his well-known essay "Signature Event Context" (Derrida 1977), sending a message relies on its fundamental capacity for displacement. The fact that a message functions after it has been sent from A to B means that it "breaks with its context" (9) and has an "iterative structure, cut off from all absolute responsibility." In other words, one can never be certain of its meaning.

Williams. The cultural critique points again to a very different aspect, one more related to the link of communication with "communion." In his *Keywords: A Vocabulary of Culture and Society,* Williams (1985, 72) discusses the force of communication that lies in its distributive act: "make common to many, impart." When communication makes something common to many, however, two very different things can happen, as Williams points out: it can "transmit" in "a one way process" or "share" (72). In this capacity,

communication has the force to manipulate as well as to
integrate and foster participation.

> **Haraway.** Not far from this position, we find the import-
> ant take of Donna Haraway on communication technolo-
> gies. In "A Cyborg Manifesto" (Haraway 1991), she points
> to a very specific force by showing that communication
> technologies create social relations that structure our
> identity, which means that they can also restructure it.
> Haraway thus points out that they can be "crucial tools
> recrafting our bodies" and that "they should also be
> viewed as instruments for enforcing meanings" (Haraway
> 1991, 164). According to her, communication can be a
> discursive weapon.

Although the preceding approaches articulate very *different* perspectives and motives, all of them notice *a force* happening when
there is communication—a force that is shaping our situation
through shaping the possibilities of communication (Shannon and
Weaver 1949), a force that can never be fully controlled (Derrida
1977) and, from a very different perspective, a force that can reach
but also manipulate the many (Williams 1985) as much as it can
be used as a weapon (Haraway 1991) to restructure our discourse.
This chapter continues their productive suspicion that communication is always more than a transparent exchange of information. By
looking into the specific case of digital technology, it explores the
hypothesis that the rise of digital media is accompanied by a specific force, which differentiates it from other technologies. To enquire
about this, it is necessary first to look into the theoretical setup of
digital media. Can such a force also be located when it comes to
digital technology?

When approaching this question, one quickly notices a rather
confusing situation. Recent studies of digital technology (Bratton
2016; Chun 2016; Crawford and Joler 2018; Gitelman 2013;
Starosielski 2015) have rightly pointed out a feature specific to
digital communication, which is shaped by a situation far more

complex than a "communication channel." Bratton (2016) has most explicitly developed this thought, showing that the technical layers of the internet's OSI architecture, by now grown into a network of planetary scale, can be described as a "stack." To explore communication, different layers of this "stack" must be taken into account: the material communication layer providing energy and matter, controlled by an optimization layer and used by an application layer (53), for example. Here network communication challenges previous theories of software.

Being written in code, software has been organized by two strands of communication and, with it, two interfaces: one for the machine (an interface whose alienness Finn Brunton explores in chapter 1 of this volume) and one for the user (an interface whose alienness I explore here). Their conflating layers are the reason why Wendy Chun (2011, 3), informed by her double degree in both systems design engineering and English literature, has called software "a notoriously difficult concept":

> Software perpetuates certain notions. . . . It does so by mimicking both ideology *and* ideology critique, by conflating executable with execution, program with process, order with action. Software, through programming languages that stem from a gendered system of command and control, disciplines its programmers and users, creating an invisible system of visibility. (Chun 2008, 316)

The disciplinary machine that software is affects programmers and users alike, as Chun points out. Following her, Alexander Galloway (2012) has addressed the interface as effect and ethos to make a similar point: interfaces do not simply transmit our messages; instead, they open—or enforce?—a very particular dialogue with technology, a point that needs to be pondered for a moment.

When discussing digital media, media theorists have often differed over where the force of digital technology originates. That there is a force, they agree—the algorithmic, as, for example, Rita Raley (2016) pointed out in her precise essay on algorithmic translation, is not purely mechanical. But where is it that media and technology scholars have to look? Do they need to look at the code with which a programmer is communicating and to which Paula Bialski turns in chapter 3? Or is the force located in the graphical user interface communicating with the user? When approaching digital technology, we too often follow "the logic of what lies beneath," as Chun (2011, 20) notes, even though "code is also not always the source, because hardware does not need software to 'do something'" (25). To make things even more complicated, further technological developments have stressed different parameters, such as data (Gitelman 2013) or machine learning architectures (Mackenzie 2017), and more parameters at the moment still unknown will follow. Thus, when looking at digital technology, this chapter assumes that for the process of communication, multiple interconnected layers are playing a part. Being interested in a very specific aspect of our dialogue with technology, however, this chapter does not focus on each of those layers but studies one particular moment: the moment when technology is addressing us. Whereas Brunton before me turns to Licklider to explore the complex setup that enables machines to communicate with each other, and Bialski in the next chapter turns to programmers to study the code review process, my chapter looks at the situation that enfolds when machines communicate with us. For this, it first needs to clarify its method of approaching technology.

As stated earlier, when technology communicates with us, it is not technology itself that raises its head and starts to speak—technology is *not an acting subject.* As Heidegger has pointed out, technology has also always been *more than an object*; that is, it has always been more than a means to an end. If it is neither a subject nor an object, however, how can in our case the force of communication regarding digital technology be approached? Here Hannah Arendt's ([1958] 1998, 151) short take on the problem of technology, which she develops while discussing the transformation of human life through technology, points our thoughts in an interesting direction:

The discussion of the whole problem of technology, that is, of the transformation of life and world through the introduction of the machine, has been strangely led astray through an all-too-exclusive concentration upon the service or disservice the machines render to men. The assumption here is that every tool and implement is primarily designed to make human life easier and human labor less painful. Their instrumentality is understood exclusively in this anthropocentric sense. But the instrumentality of tools and implements is *much more closely related to the object it is designed to produce.* (emphasis added)

Here Arendt states that any given technology is more closely related to another technology than to a human subject. To her, technology is driven by an immanent ("closer") relation. This does not mean, however, that technology acts as a subject that masters the human. Humans play a part in the development of technology, which becomes clear in an "important assumption" added by Arendt: "that the things of the world around us should depend upon human design and be built in accordance with human standards of either utility or beauty" (152). Pleading for human standards, Arendt shifts the focus onto technology in an interesting way. She approaches it more *as a situation* and less *as a subject,* which becomes explicit in the following quotation: "The question . . . is not so much whether we are the masters or the slaves of our machines, but whether machines still serve the world and its things" (151). This chapter follows her approach when studying the force of communication by investigating how technology as a situation can be thought of in more detail. What should be examined? How does a technical situation need to be studied? To answer these questions, the chapter links Arendt's approach to Gilbert Simondon, with whom her take on technology resonates.

Like Arendt, Simondon (2017) finds our understanding of technology fundamentally flawed. Instead of emphasizing curiosity or understanding, Simondon critically remarks that our usual approaches toward technology oppose humans and machines (15). To overcome this, he rethinks this relation. In the chapter "Evolution of Technical Reality: Element, Individual, Ensemble," he describes how technical evolution is not driven by men or machine but by an "ensemble" of the two. There is no master anymore who is in control of the process of a technical development. And this shift from a master relationship to an ensemble raises a question: instead of a gifted inventor or mad genius, what drives the development of technology?

For Simondon, similar to Arendt, the answer lies in the productive relations between men and technology, which create a process of "concretisation" (Simondon 2017, 33; also Iliades 2015). He sees this, for example, in the development of X-ray tubes: regarding the Crooks tube and its later "successor," the Coolidge tube, Simondon finds the engineer William Coolidge elaborating on technical functions of the already existing Crooks tube. Coolidge "purified" them to improve the tube's functioning—a process of concretizations in which specific aspects of an already existing technology get further developed: "the functions are thus purified by their dissociation, and the corresponding structures are more distinct and richer" (36). Instead of being struck by a flash of genius, it is the "technical reality" of the Crook tube that inspires the new product. Thus it is the technical reality itself that fosters further development, although this reality needs the human to concretize: "machines can neither think nor experience [*vivre*] their mutual relation; they can only act upon one another in actuality, according to causal schemes." With this, the role of the human comes into play: "Man as witness to machines is responsible for their relation" (157).

Neither human nor technology can initiate the process of further development on its own. They need to relate to each other. With the human as an *enabling witness,* the relation of man and machine can be sketched as an *ensemble* instead of as an opposition. This puts the human in a very distinct role: the human is *not master* of machines digital or mechanic but their *interpreter.* In Simondon's (2017, 150) words, "man understands machines; for there to be a

true technical ensemble man has to play a functional role between
machines rather than above them" (see also Combes 2013, 57).
Here the concrete technical relation of a technical object to its
milieu describes an immanent development driven by "concreti-
sations" that are nondirectional. Fascinated by constant technical
change, Simondon (2012, 13) will later describe technology as characterized by an "*opening*": "technical reality lends itself remarkably
well to being continued, completed, perfected, extended." Thus, in
the middle of this, one finds an interesting tension: technology puts
forth a situation that then needs a human to continue, complete,
perfect, and extend it, in short, to turn it into reality. At the same
time, technology follows its own, alien logic in what it offers to be
continued, completed, perfected, and extended. We cannot predict
the future of the technology we have invented. Even in the twenty-
first century, in which we are facing a field as closely guarded as an
economy driven by digital technology, we are never certain which
technology will become the "next big thing."

Technology is a force alien to us that has now started to speak
and process language. But just because it has started to process
language and can now say something, we should not mistake it
for a speaker. *Being with technology* instead means to approach
technology as a technological ensemble, as a continuously
developing situation made up of humans and technology. Thus
we need to study what kind of situation unfolds when technology
communicates with us as we aim to avoid treating technology as
an anthropocentric subject that acts and/or speaks. Luckily, a blue-
print for the power of communication that does not stem from a
subject (although a subject is involved) can be found in the concept
of interpellation Louis Althusser introduces when discussing the
notion of ideology.

Althusser's notion of ideology evolves around an interesting shift.
While he analyzes communication (or interpellation), he does not
look at what is said or what can be said. Instead, Althusser (2014)
focuses on the situation created when being addressed and the
force of this address. In his essay "Ideology and Ideological State

Apparatuses," he analyzes the structural force happening in the moment of communication. Using the example of a policeman calling out to you on the street, he illustrates that communication situates (even appropriates) its participants by establishing a link between sender and receiver in the act of interpellation: it constitutes a subject. His description of this constitution has turned into a highly influential theory of interpellation, although it is less a "theory" than just a few paragraphs. In those paragraphs, Althusser shows that a specific social role—in his words, a "subject"—comes into being by "the practical telecommunication of hailings" (264). To illustrate how this "hailing" or "interpellation" functions in the context of ideology Althusser introduces an individual that turns around in response to a policeman shouting "Hey, you there!" (264) to "answer" that call. And in exactly that moment, so Althusser, one becomes a subject relative to the ideology of law and crime. In other words, in that moment, one experiences the social force of communication, which Althusser calls ideology: "ideology 'acts' or 'functions' in such a way that it 'recruits' subjects . . . . , or 'transforms' the individuals into subjects . . . by that very precise operation which I have called interpellation or hailing" (264).

In the twenty-first century, this operation of interpellation Althusser described, an operation that creates a situation of recruitment by establishing a link between a sender and receiver, is still continuing. Only now, it can be found in new and different forms of communication—and this is the hypothesis I would like to bring to a test in this chapter: Today, the recruiting of subjects happens when technology addresses us. By interacting with the interfaces of technology, we are situated through this communication and recruited as specific subjects. Of course, that we make a world for others to live in through our technological creations has been an aspect in philosophy of technology, which Langdon Winner (1986, 17) but also Donna Haraway (1997) and many others have addressed in much detail. This chapter adds to those explorations of politics we built into our technologies, although it will be slightly shifting the view. By approaching technology with Arendt as a situation and by trying to understand the contemporary technological ensemble (Simondon), it will not look at what is being said to us by technology. Instead, it is interested in the kind of situation that unfolds. As what kind of subject are we recruited in that situation? The next section therefore observes the communication with technology to tune into how something is being said when technology addresses us.

How Is Technology Addressing Us?

To capture how technology addresses us, this section analyzes three different examples partly drawing on earlier research (Bunz 2015): it looks at the introduction of Apple's iPad to study its early interface design, considers the brand communication of internet companies and their fondness of mascots, and, finally, turns to the Google Doodles that appear on the landing page of Google search, which one passes by when searching for other information.

On April 3, 2010, Apple's cofounder, chairman, and chief executive officer unveiled a tablet computer it introduced as "iPad." Its new product was operated via a touch screen and could play music, take photos, shoot video, and perform internet functions such as web browsing and emailing; more applications, from games to social networking, could be added. In its first fiscal year following the launch of the new product range, Apple sold 32 million iPads, with 140,000 apps being created for it by December 2011 (*Economist* 2011). One could say that with the success of the iPad, a new era in the relationship between human and computer materialized: the tablet computer showed that digital communication had left the workplace to become a commodity in our day-to-day lives. Computers had certainly entered leisure time with game consoles long before. The iPad, however, could be used for much more than just gaming. It could perform all tasks done by a personal office computer at that time, although it was not supposed for working. Its reduction to a large touch screen that weighed 680 grams made it comparable to a heavy book or magazine that could be read at home on the couch. It was its slick materiality that differentiated it from a computer as much as its specific user interface.

By that time, screens had been technically refined so that their visual interfaces no longer needed to be operated via minimal black-and-white icons. They could be replaced by touch screens with voluptuous 3D buttons more to the taste of Steve Jobs. As the former CEO of the animated film studio Pixar, he had a passion for reality imitating 3D graphics, as had Scott Forstall, the first architect of iOS, the software developed for the iPhone and iPad. Thus the early iPads had many 3D buttons and other skeuomorphic features each mimicking an original: the Notepad app had a border of stitched leather to make it look like a real notebook, the Podcasts app displayed a reel-to-reel tape deck when one pressed play, and the calendar and contacts apps looked like small books and featured a page-turn animation. Making apps and items mimic their real-world counterparts gave the iPad a stuffy look and feel. This continued in a different way Apple's traditional appeal to nontechnical people. Right from the start, the company had established its computer as a fun-to-work-on machine by including features such as greeting users with a "happy Mac" when starting or by using symbols like the "dogcow" (indicating the setup of a page), scissors (for the cut command), or the trash can, which were created by Susan Kare for the back then still limited black-and-white screens. Now computers had entered a new, advanced, but also more serious era—at least that was the impression Apple gave with their design of the first iPad. Its look and feel communicated to the user that computers had come of age, although not for very long. Technically, all screens from phones to tablets to laptops to PCs were able to display complex, grown-up 3D interfaces. Still, a new and very different trend emerged that soon became more successful than mimetic skeuomorphism.

Surprisingly, the new trend was initiated by Apple's rival Microsoft, which, after the iPhone's success, had already been written off. Faced with the staggering success of Apple's phone, Microsoft had to respond with an original and different approach: for their handheld devices, the Microsoft designers decided to focus on cards and not on buttons. Eager to avoid Apple's extensive use of skeuomorphism (Wingfield 2012), their inspirations came from the design principles of classic Swiss graphic design, which favors a minimal style, emphasizes typography, and uses a grid that can often be seen on European transportation signs. Instead of buttons, they used text placed on cards, which one could navigate laterally through scrolling canvases. Their typography-based design language came to be known as Microsoft design language. Its principles had originally been developed for Microsoft's mobile media player Zune (2006–8), before they were taken over to the Windows phone, launched in 2010. Although the device did not have the same success as the iPhone, its design would inspire others, Google among them—and Google's logo in fact exemplifies this new and different approach to user communication.

While Apple's skeuomorphic design for the iPad communicated its device as a toy-tool for grown-ups, the flat design Microsoft had initiated would go a very different way—and with it a new form of addressing the user would begin. Early on, Google would be part of this. On Wednesday, May 5, 2010, the search engine Google changed its logo for the first time in ten years and eleven months (Googleblog 2010). The new logo was less skeuomorphic and more colorful. Its three-dimensional letters in red, yellow, and blue, plus the green letter *l* based on the font Catull, lost their drop shadows. The logo had exchanged the rich details of skeuomorphism in their big typography with louder colors and simpler forms. Google's senior user experience designer Wiley explained the change on the search engine's blog as follows: "The new logo is lighter, brighter and simpler. We took the very best qualities of our design—personality and playfulness—and distilled them" (Googleblog 2010). Experts agreed. Already before the change, British graphic designer Peter Saville, known for minimal design like the radio signal cover for Joy Division's album *Unknown Pleasures,* described Google's logo in an interview not just as playful. For him, it was addressing children: "Everything about it is childlike: the colors, the typeface, even the name" (cited in Rawsthorn 2010).

The redesign intensified this further. Chris Moran, then the *Guardian*'s search engine editorial optimizer, commented on the new look and feel as a turn toward "My First Search Engine" (pers. comm., May 6, 2010). Online, the rise of flat design had begun, even though it would take a while before its triumph over skeuomorphism became recognizable—it was not until 2013 that an animated web page displayed the "battle flat design vs. realism" (Intacto 2013). Flat design opposed skeuomorphic and other "artificial" design techniques in favor of two-dimensional, "flat" illustrations; big typography; and bright colors for a more simplified aesthetic. When the new design became a mainstream trend, however, something else changed—technology would approach the user in a different way. The new design style addressed a very different user—not an adult one. Visually, the style resembled books for very young children. Addressing the user as a very young child, however, was a transformation that did not happen abruptly and not just in one field. With hindsight, years before 2013, the new trend in brand design could have been spotted on the World Wide Web. And although it went unnoticed for a long time, it fundamentally changed how brands approached the user.

Contemporary brand communication generally has a double function: it enables the user to identify a product and, for this, gives the product or service a specific identity or image (Millman 2012; Holt 2004). With the internet, as many marketing books were eager to explain (Levine et al. 2000), brands had to become a conversation. But this was not the only novelty. Online, the rules seemed to be different, which is why several internet companies embraced animals (or aliens). Or was it because they addressed *someone* very different? In any case, if one attentively observed the brand communication of "online" products and services, one could notice that animals had peacefully appeared in large numbers. Next to the fox of the web browser Firefox chirped the blue bird of the microblogging service Twitter, while a little white alien with antennae accompanied Reddit, a social networking service that provided online conversations for "digital natives," as they were dubbed. And not only platforms but also technology companies seemed to have a thing for mascots, from Tux, the penguin of the Linux operating system, to the black Octocat that had landed on the 404 pages of Github, the web-based hosting service for software development projects. And there were many more, like the bare-bellied chimpanzee with a postman's hat who helped create professional email for MailChimp; or the big-eyed brown owl that had become part of the logo of Hootsuite, a social media management dashboard; or the flying beaver that sat enthroned on the online travel page of a start-up company called Hipmunk. Even a nonmascot service like Facebook introduced a character, the Zuckasaurus, which looks "like a short Barney, the kid's television show dinosaur" (Bilton 2014). Standing on its two feet while checking its laptop, the blue dragonlike dinosaur was first spotted in April 2014, when it started to address users in a pop-up window with the educational concern that it "just wants to make sure you're sharing this post with the right people" (Bilton 2014). In short, animated animals could be found all over the World Wide Web as if it were a fairy tale. Mascots had spread from sports, where they were supposed to bring luck to a team, to the internet, and academic books started to analyze the phenomenon (Brown and Ponsonby-McCabe 2014). In the offline world, brands that were targeting their products to adults generally refrained from using mascots; companies that produced cars, alcohol, or even entertainment electronics rarely considered an animated animal as part of their brand strategy.

Parallel to the appearance of the online mascots, a similar development could be found on search pages: the rise of the Google Doodles, which introduced a new, unique style of commemoration that shared the same tendency. Until 2010, Google had only sporadically changed its prominent search website logo into those "Doodles" to mark an anniversary or event. Although the concept of the Doodle was born at the very start of the company (1998), when founders Larry Page and Sergey Brin changed the logo with a stick figure drawing to mark their visit of the Burning Man festival in the Nevada desert, the logo was not changed very often. It took two years before they requested a second change to honor Bastille Day, commemorating the beginning of the French Revolution each year on July 14. Before 2010, the logo was changed only on rare occasions. Then one could find a sketch that playfully intertwined the topic of an event with the logo: the birthday of English mathematician Ada Lovelace, Martin Luther King Jr. Day, or Halloween. After 2010, the frequency with which Doodles replaced the logo intensified. In 2010, Google published thirty-five Google Doodles, more than in any previous year. In the years 2011 and 2012, this number went up to seventy-six and eighty-three, respectively, and has gone up ever since. More and more Doodles displayed events or presented persons shaping human history and culture with imaginative cuteness. They started to appear worldwide, thereby taking national cultures into account: Britain celebrated the eight-hundredth anniversary of the Magna Carta (2015), Mexico the Day of the Dead (2013), and the United States the Mexican Hollywood actress Katy Jurado (2018).

Considering that Google is now an essential part of our public sphere—the Court of Justice of the European Union (2014) indicated this by its ruling that natural persons have the right to be forgotten and links to personal data must be erased in this public space—Google Doodles are the monuments we find in it. As we pass by those monuments when searching, we are reminded of important moments that have shaped our human fate. This form of commemoration, however, happens in a rather unique way, different from historic monuments cast in stone and erected on our public squares, which foster a certain symbolism and spread an air of pathos. Indeed, most public monuments in stone or bronze are slightly pathetic, from the Statue of Liberty enlightening the world from Liberty Island in Manhattan to the Soviet War Memorial in Berlin's Treptower Park to the Monument of the People's Heroes in Beijing's Tiananmen Square to Christ the Redeemer in Rio de Janeiro cresting Corcovado mountain. Online Doodle monuments, on the other hand, turn achievements into playful stories with imaginative cuteness and are supposed to be "fun" (Google Doodles Archive 2018). It should come as no surprise that they more often commemorate birthdays than deaths.

Before judging Google Doodles as "history light," however, it is important to take a step back and get a full view of the transformation. Certainly all three developments—the rises of flat design, brand mascots, and Google Doodles—show a common tendency, as their style is equally defined by colorful surfaces, big typography, and playful stories or mascots, thereby resembling elements we are familiar with from children's books or apps. Thus what is the specific form of interpellation that can be noticed here? How is technology addressing us? To state the obvious, online technology has started to address us as if we were children. The extent of this infantilization, however, only comes fully into view when comparing the described design tendency to an older project designed by Dieter Rams, who helped the company Braun to relaunch an educational toy called Lectron; and like many of his other designs, it became iconic.

Lectron was a modular electronic experimentation kit designed to introduce youth to basic electronic circuits and theory. From 1967 on, the German designer and his team, among them Jürgen Greubel, produced the packaging in a new style, including a redesign of all manuals. Being supervised by Rams, it is not very surprising that the Braun Lectron Hobby Set Radio Receiver (1969) is kept in a minimal style. Contrary to the users of Google's search engine, Apple's iPad, or the service online brands, however, it does not target adult users. As a game, it is tailored to a much younger age group. So how does Lectron approach its teenage user?

The cardboard box cover shows three photographs. Two smaller ones display the white radio set in Rams's minimal design and a detail of a printed circuit board; the bigger photo pictures a black-haired teenager in a buttoned-up blue shirt, who sits in front of components and tools soldering electric parts. Lectron approaches

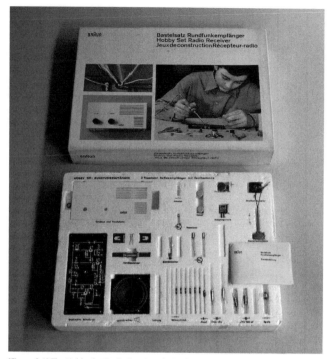

[Figure 2.1] The Hobby Set Radio Receiver design by Dieter Rams and Jürgen Greubel, 1967. Photograph by dasprogramm.

the technically interested and capable teenager. Contemporary flat design, on the other hand, incorporates design elements for a much younger age group. Its colorful surfaces, big typography, and animated characters are generally design elements used for targeting children aged two to seven—a time during which children are in the sensorimotor stage. Children in this stage, as the child psychologist Jean Piaget has shown, assign active roles to things in their environment (animism), while their activities are mainly categorized by symbolic play and manipulating symbols. It is a stage in which physical operations are more dominant than mere "mental" operations. Thus the conclusion is obvious: we are addressed by technology as very young children.

Fighting back the natural reaction to all miscategorizations (feeling insulted), this is an interesting outcome to be investigated further by shifting our attention back to the aspect Althusser had in mind when discussing being addressed as a form of power. So what is the effect of this infantilization of user interfaces? What force or form of power play are we facing here? For that we face a form of power play can almost be taken for granted—when technology is communicating with us in this way, it is surely not just transmitting the friendliness of cuddly Silicon Valley companies that commissioned plush toy–like interfaces to comfort us in the exhausting world we live in. To understand this manipulation further, the next section categorizes this infantilization.

How We Are Getting Manipulated

Technology has always manipulated us (Winner 1989, 19), and it does this more openly than ever, since it has started to speak. For this, one does not even need to turn to conversational interfaces, such as Apple's Siri or Amazon's Alexa, quarreling with us if the lights should be on or off. This also can be easily noticed by anyone who has been disciplined by a car's navigation system. In fact, Global Positioning System (GPS) usage is a good example of a simple form of manipulation, as it has turned into quite a dominant system. To get their exact position, smartphones and millions of other devices use GPS, which was launched 1978 by the U.S. government. The system's Master Control Station is located in the Schriever Air Force Base near Colorado Springs, overseeing thirty-two GPS satellites (U.S. Naval Observatory 2018). Currently only Russia operates an alternative system, GLONASS, with Europe and China working on further alternatives. But most cars and smartphone maps use the GPS signal, which is then correlated to a road or a calculated route. The route, however, does not always coincide with reality. A survey for Michelin (2013) among 2,200 U.S. drivers showed that 63 percent of those who use GPS say that it has led them astray at least once by pointing them in the wrong direction—and some of us obey those directions more than others.

In the United Kingdom, a driver continued to follow the navi's instructions, which told him the narrow, steep path he was driving on in Todmorden, West Yorkshire, was a road. He only noticed the mistake after he struck a fence and his BMW hung off the edge of a cliff. In South Brunswick, New Jersey, a driver ignored the end of a road because it was differently displayed on his navigation system. Following the navi's version of reality, he ignored a stop sign and hit a house. In Australia, three Japanese tourists drove their car into the Pacific Ocean. Their navi had told them there was a road to the North Stradbroke Island. After five hundred meters, they got stuck in the mud, their car being flooded by the tide. In Bergün, Switzerland, the navigation system told a man to turn onto a trail. The trail was for goats. The minivan that he had driven up that trail could only reach the road again with the help of a heavy-lift helicopter. In Italy, two Swedish tourists drove four hundred miles to the wrong Capri. Instead of relaxing on the island with its blue grotto, they ended up in an industrial city in Italy's northern region that bears the same name. In all cases, human judgment was distorted by technology, it seems. But the dialogue between human drivers and advising technology only looks at first sight like a master discourse, in which human servants blindly follow a directing technology. Technology, as both Simondon and Arendt have reminded us, is not necessarily an opposing force that aims to bring humans under control and is wrongly thought of through the template of master and servant. After all, in the preceding cases, the advice of technology could have easily been ignored. Thus one could also say that in most cases, the drivers, often tourists who were not familiar with their environment, followed "their" technology instead of asking other humans for help. In other words, we are part of this manipulation—and the same is the case when we look at patronizing, talkative self-service checkouts.

One of the countries in the West that embraced self-service check-outs early was the United Kingdom. By 2015, Tesco, the United Kingdom's largest supermarket chain, had already introduced twelve thousand of them. To help shoppers understand how to operate the new technology, the checkouts give verbal guidance on how to use them. And their most renowned comment in their early phase became "Unexpected item in bagging area. Remove this item before continuing." The reason for this comment: its pay mechanism has integrated scales. It weighs the item after it is placed in the grocery bag; this is done to ensure that the shopper pays for all the items in the basket. The problem is, however, that the system gets easily irritated, for example, when an item is too light and the second scale fails to recognize it. In these cases, the checkout announces loudly that there is an "unexpected item in bagging area" and soon after starts nervously flashing a light and an alarm sound for everyone to hear and see—the system calls for help, as it needs the reassurance of an assistant. Does it accuse you of being too thick to use it? Or suspect you of being a thief who has just stolen something? Being addressed by it in an Althusserian manner—"Hey you, there!"—we react annoyed. We recognize that other humans who see and hear this might put us into the category of social subjects who have problems using a self-service checkout, which is not very flattering.

Here we experience manipulation: when making you behave in the right manner or advising you to do the right thing, both the self-checkout and the car navigation assistant are forms of disciplinary manipulation, in contrast to those open forms of manipulation we find with infantilization, which do not directly tell you what to do. This seems to be of a different kind, with its interface *not* disciplining us but simply suggesting a situation. Cheerful design signals a simple and unproblematic context. By addressing us as very young children, the playful interfaces of flat design suggest that there is no need to understand anything. Just try it: go press this button, speak to it, create! The simple but colorful appearance signals that the users can be free from second thoughts about the complexity of the technological apparatuses as well as about the complexity of the world we live in.

We are manipulated into a situation we seemingly don't have to question—and this is why we should pause. For we have reached our first conclusion: having looked at how technology is addressing us, this chapter could establish that it is recruiting us as very young children. But can we really read the situation as technology concealing its mode of operation to lure us into its unquestioned usage? Would this not mean that we have positioned ourselves again in opposition to technology? After all, this chapter does not plan to study the concealed interests of technology companies. Instead, it aims to analyze and understand our *being with technology* by analyzing our current dialogue with it through looking into its actual "concretization" (Simondon 2017); indeed, Simondon discussed the intuitive approach of children toward technology as one way of understanding the being of technology: "One cannot study the status of the technical object in a civilization without taking into account the difference between the relation of this object to the adult and to the child," he writes (106). The technical training of the child is based on practicing with technology bringing forth a "technical subconscious" (107), which can also be understood as an intuitive skill. This experimental skill is a certain intuitive mode of technical knowledge also linked to "experts"; Simondon names the operational knowledge of farmers or of craftsmen about the material they work with. Their technical training consists of "intuition and purely operative schemas that are very difficult to formulate or transmit through any kind of symbolism" (107). Instead of scientific knowledge, the operational knowledge is created through technical realization:

> Technical realization, on the contrary, provides the scientific knowledge that serves as its principle of functioning, in the form of a dynamic intuition that can even be apprehended by the young child, and which is susceptible to becoming more and more elucidated, doubled by a

discursive form of comprehension. . . . Through technics, encylopedism could thus find its place in the education of the child without requiring capacities for abstraction, which the young child does not fully have at its disposal. In this sense, the child's acquisition of technological knowledge can initiate an intuitive encyclopedism, grasped through the nature of the technical object. (124)

Following Simondon, and linking his understanding of intuitive encyclopedism to our problem if being recruited as very young children, one could therefore also understand the "call" of technology as an invitation to learn about a digital interface. We, however, read this dialogue according to the idea that technology is manipulating us into being its slave users, which seems to be a rather anthropomorphic reading of technology: it treats technology as if it were a human in the role of an acting subject. As pointed out earlier, technology has agency and is a force, but to understand the alienness of this force means to remind ourselves that it is not a human subject that follows a Hegelian interest to subjugate and control other humans.[1] Technology creates specific situations—in this we can find its force—but when creating those situations, it does not follow a specific interest, and this is exactly why Donna Haraway (1991, 161) in "A Cyborg Manifesto" sees the potential for "rearrangements in world-wide social relations tied to science and technology." What is created by technology can always be interpreted in different ways—if its force is understood. Even Marcuse (1998, 42), whose take on technology is generally rather critical, writes that "technics by itself can promote authoritarianism as well as liberty, scarcity as well as abundance, the extension as well as the abolition of toil." Technology is not neutral—its force is that it confronts us with a specific situation or a specific transformation; how this transformation is interpreted, however, and which concretization is going to appear is always adapted by us humans, as we are part of the technical ensemble. To say it with Donna Haraway: "We're living in a world of connections—and it matters which ones get made and unmade" (cited in Kunzru 1997).

Returning with this insight (that technology creates situations, although without interest) to our childish dialogue with technology, reading this dialogue through Simondon's approach of an intuitive encyclopedism, we can still find a negative effect of our infantilization: the creation of a situation that does not need to be further questioned. But can the recruitment of technology addressing us in an infantilizing manner be thought of differently? Can we move beyond the template of master and servant? To follow this question, the next section explores infantilization from a different perspective, by looking at an advertisement of the company that created the style of flat design: Microsoft.

In 2014, Microsoft aired its first national Super Bowl advertisement, a one-minute video produced mainly in-house. Using Microsoft products, it explores technology through the eyes of Steve Gleason, a former NFL player who is battling ASL, a severe illness that attacks nerve cells in the brain and spinal cord that control muscle movement. At the beginning of the video, we hear a computer-generated voice asking, "What is technology?" and see it being written by Steve Gleason, who sits in a wheelchair with a keyboard he operates via eye movements. We see a girl playing with a red windmill. From there, the commercial cuts to symbols that resemble written code, followed by Microsoft's colorful card screen design. Then a surgeon is flipping through large medical images displayed on a wall using hand gestures, followed by a white toy robot, which is about to look at us, as the camera movement suggests. Gleason's next question can be seen and heard: "What can it do?" after which a small boy enters the screen playing baseball standing on two artificial legs, followed by the ninety-eight-year-old painter Hal Lasko, partially blind, painting a colorful landscape with the help of a mouse. Again, Gleason's artificial voice is asking, "How far can we go?" We see pictures of a satellite in the universe, a surgeon using his hand to control an X-ray, and two groups of children cheering each other via a video-chat projection. After this introductory period, the next thirty seconds are grouped around a theme showing the examples of the "power" of technology, as

Gleason puts it: a soldier being remotely present during the birth of his child; a small child freaking out with joy when she sees her dad on the screen; several scientific and medical successes, from the launch of a rocket to a man with an artificial arm moving his hand and the emotional reaction of a women making remote contact with someone on the other side of the screen. It ends with the slogan "It has given voice to the voiceless," showing Gleason in his high-technology wheelchair, a computer helping him communicate, his son on his lap, to whom he now connects directly by raising his eyebrows. The main slogan appears—"Empowering us all"—to be replaced after a few seconds by Microsoft's logo.

The commercial is informed by the topic that frames it—how technology helps, "empowers," those we love and care for to lead better lives—and certainly appeals to our emotions. The majority of the situations depicted in this video are related to health and science. Thus the situations visualized mainly pertain to health or science—generally areas not dominated by children. The video, however, uses nearly as many images of children (as individuals and in groups) as of adults. A content analysis[2] shows nine sequences with the focus on children and twelve with the focus on adults. The reason for images of curious, excited, and playful children lies partly in the task of every commercial: to create appealing images. But there is more to it. That children are playfully discovering technology is also symbolic. This becomes apparent when Gleason's first question opening the video—"What is technology?"—is followed by a sequence showing a small girl in a dress curiously looking at the windmill she puts into motion with her small hand: humans exploring technology. The message of a girl putting a windmill into play (its movement enhanced by a sound effect) is visually answering this question. Moving a windmill means exploring technology. *The usage itself* is an act of exploration—and empowering.

Of course, one can argue that this is a message in the interest of Microsoft: the sheer usage of its commercial products is empowering—and not programming code yourself, as, for example, open source software would allow. Being able to understand or even program code yourself can certainly be more empowering. Still, this does not fully explain why the question "What is technology?" finds a fitting visual sequence in a child playing with a windmill. Instead of asking what a windmill has to do with digital media or Microsoft, the sequence makes sense. Linking this image to theories of learning and its role for the history of graphic user interfaces, the next section aims to explain why this could be the case.

Logic Is Not a Derivative of Language

The graphical user interface has become a commercial success, although this took several experiments, among them Douglas Engelbart's NLS system, Ivan Sutherland's Sketchpad, SGI's Iris, the two interfaces of the Xerox Alto and Xerox Star, and the Apple Lisa and Apple Macintosh. As such, it is generally referred to as the transformation that helped personal computers to become mainstream (e.g., Chun 2011, 59). Its advantage: it is easier to use than a command line interface. Therefore the graphic interface appeals to users not familiar with coding. This section aims to inquire what it is that makes it easier and how this is linked to the girl playing with a windmill. To show this, it is first necessary to compare the older command line interface with the newer graphical user interface with respect to learning. In principle, both interfaces have the same function: they are ways to command a program. How they approach the user, however, is different. A graphical user interface's windows, icons, menus, and pointer are intuitive elements, whereas the knowledge to operate the command line needs *to be learned beforehand.* A graphical user interface can be operated without much knowledge as it *incorporates the learning into its usage.* Learning theories in fact played an important role in its development. Discussing the work of mathematician Seymour Papert (1963, 1968), who collaborated closely with child psychologist Jean Piaget and also influenced the computer scientist Alan Kay, this section takes a look at the connection of learning theories to computer science in general and the graphical user interface in particular.

When developing new approaches to artificial intelligence, Papert had come across theories of learning by child psychologist Jean Piaget. The South African had met Piaget when he spent time in Paris as part of his second doctorate in St. John's College in Cambridge and decided to follow him to his Institute in Geneva to apply his theories to artificial intelligence, a field that found itself in its golden years from 1956 to 1974, driven by new discoveries and funding. More precisely, Papert's aim was to enhance machine learning by incorporating Piaget's ideas of the learning of children, although their interest was mutual: Piaget endorsed Papert's cybernetic approach and published many of his articles in his journal *Études d'Épistemologie Génétique.* Known today as a child psychologist, he understood himself as a scholar of epistemology exploring theories of knowledge with the aim to establish a new approach toward understanding. And it would be the graphical user interface that would pick up this approach to show that children's learning can indeed be applied to adults' learning too.

Interested in multiple ways of knowing, Piaget turned to children's learning as a unique form of interacting and theorizing. Curious about their thinking, he took their logical reasoning seriously, even when their thinking led to "wrong" answers. His nonjudgmental approach enabled him to describe four universal stages of cognitive development that are still relevant to contemporary psychology. More important in the context of this argument, however, is something different: central to his approach was the hypothesis that for human understanding and learning, the act of reasoning (the work of the mind) is as important as practical or experimental understanding (the work of the fingers and mind together). When observing children between the ages of two and seven, Piaget recognized a specific way in which children play. He saw in children's sensorimotor approach a form of learning—thinking with fingers—most important when we are very young children. From this, he concluded that logic is formed not only in the brain:

> I believe that logic is not a derivative of language. The
> source of logic is much more profound. It is the total

coordination of actions, actions of joining things together, or ordering things, etc. This is what logical–mathematical experience is. (Piaget 1972, 13; see also Piaget 1969, 90)

Piaget developed what has come to be known as constructivism, an approach that viewed learning as a reconstruction rather than as a transmission of knowledge. It valued experience highly and understood playing—the manipulating of materials—as a way to create knowledge:

> To know an object, to know an event, is not simply to make a mental copy, or image, of it. To know an object is to act on it. To know is to modify, to transform the object, and to understand the process of this transformation, and as a consequence to understand the way the object is constructed. . . . In other words, it is a set of actions modifying the object, and enabling the knower to get at the structures of the transformation. (Piaget 1972, 20)

To apply and automate this approach to machine learning, Papert (1963) developed a project called "genetron," which explored the learning of algorithms by allowing them to build their own network topologies that simulated qualitative and quantitative developmental change (Shultz et al. 2008; Minsky and Papert 1969). He was later assisted by Marvin Minsky, with whom he cofounded MIT's Artificial Intelligence Lab. Despite support from MIT, the project struggled with technical limitations (Shultz et al. 2008). But Papert had also started to approach the relation of child and machine through another angle, manipulating not the machine's learning but children's learning. Applying Piaget's theory, the aim here was to allow a coordination of actions—acting with an object—to initiate learning in children: learning to operate a computer. Together with his colleagues Wally Feurzig and Cynthia Solomon, Papert developed LOGO, an educational dialect of the functional programming language Lisp, which was used to command first a virtual turtle, then a small turtle-shaped robot that could move and draw. And it was this approach that would inspire Papert's colleague

Alan Kay (1972) to develop a graphical user interface not just for children but also for "children of all ages."

When he met Papert, Alan Kay was a young, creative computer scientist who had thought about the graphical user interface ever since he was a student—the first thing his supervisor gave him to read was Ivan Sutherland's description of the Sketchpad, one of the first interactive computer graphics programs. But it was watching children in schools using Papert's LOGO that enabled a breakthrough:

> Here were children doing real programming with a specially designed language and environment. . . . This encounter finally hit me with what the destiny of personal computing really was going to be. Not a personal dynamic vehicle, as in Engelbart's metaphor opposed to the IBM "railroads," but something much more profound: a personal dynamic medium. With a vehicle one could wait until high school and give "drivers ed," but if it was a medium, *it had to extend into the world of childhood.* (Kay 1996, 523, emphasis added)

Kay understood that the logic of the world of childhood could be extended to adults by reapplying visual thinking to an adult interface. Reading (besides Piaget) the educationalists Jerome Bruner and Maria Montessori had convinced him that not the command line but visual thinking and a more iconic approach (531–32) would shape future ways of operating a computer. His insights culminated in his proposal "A Personal Computer for Children of All Ages" (Kay 1972), which described a portable educational computer to be commanded by experimental actions. It was based on a program that came to be known as Smalltalk, a program "environment in which users learn by doing" (547). Via Papert, Piaget's insight that logic can be a coordination of actions had found its way to Kay's interface; Kay saw Piaget's thesis confirmed: "Just *doing* seems to help" (547)—a seismic shift. With the graphical user interface, experimental thinking started to assist linguistic thinking. And with the rise of digital media, interfaces have become the way we approach information, an approach based on experimental as much as on linguistic logic. Relying on a logic we use in Western culture primarily when we are very young, interfaces address us as very young children. Users of graphical interfaces are asked to apply an experimental logic, which means to learn to understand the interface via a set of actions. Ever since the rise of digital media, the devices that inhabit our kitchens or gardens have stopped asking us to read through the manual before being switched on for the first time.

The infantilization of interfaces does not necessarily mean that technology is becoming smart while we are declared stupid. The manipulative dialogue of today's interfaces is not necessarily an act to deceive the user. Reaching out to a human logic mostly used in childhood, similar to the way Kay's and Papert's interfaces functioned, the playful addressing of the user can also be read as an invitation to experiment. In experimenting, in playing with the windmill, we use digital technology. Using it, however, means to understand how to act on it—acquire the skill to use its force— thereby entering into a dialogue with that technology. Entering into this dialogue is important not just for the case of the graphical user interface but also for artificial intelligence and machine learning, about which Shan Carter and Michael Niessen (2017) have argued that its new form of computing must be linked to a new and different interface to fully unfold its operational knowledge. To bring forth this operational knowledge in a more general sense, digital technology is calling upon us as children. It is not addressing us as adults, as engineers. To call into action an intuitive, visual-operational knowledge, marginalized in our postindustrial Western societies, it is recruiting us as children of all ages. The force of communication we face in digital technology is an operational knowledge; to make use of it, we are being framed as very young children.

The hypothesis that digital technology finds itself linked to a specific force could be shown; still the analysis cannot stop here. For within this force, an interesting setup of power relations unfolds, power relations that are coming into action when we communicate using digital interfaces. Is the infantilization of interfaces inviting us to experiment with those interfaces, or is it luring us into a playful situation that is not to be intellectually questioned? To understand our contemporary being with technology, another effort needs to be made to explore the lines of power that run through it. How do we know if a digital interface is addressing us with the aim of empowerment, or deceiving and sedating us? How can one conceive the difference? This is the difficulty when it comes to being addressed as children: the infantilization of interfaces is able to be both patronizing *and* empowering simultaneously—the power we find within the force of communication refrains from following a well-behaved dialectical thinking.

Being patronizing *and* empowering means that one cannot be *for* or *against* infantilization. Being *for* the user's emancipation does not equal being *against* infantilization. The conceptual architecture we find at work here does not unfold in an oppositional way. An interface can be both patronizing and empowering in the same moment and is therefore not fitting into the antagonistic concept of dialectics, thesis and antithesis. Questioning the phenomenon of the infantilization of interfaces further with regard to the powers at play here, however, one also can realize that at the same time, an antagonistic, dialectic relation is not completely gone: an interface can be patronizing and empowering at the same time, although to be patronizing and to be empowering remain fundamentally different acts of power. While empowering users means that we are learning to use the power of technologies ourselves, patronizing guides and shoves us toward just acting out that power. One time the power is with the user; the other time the power is just lent to the user—in other words, there is still a fundamentally dialectic relation between. Deep inside the conceptual architecture, a negative relation, this complex force of negativity that has been described by Susan Coole (2002) and Benjamin Noys (2010) for thinking/acting difference is still at play, ensuring that there is difference.

From this follows that, again, we need to try coming to grips with the force of communication and the forms of power we find in its act of infantilizing the user. For this, the last section of this text turns to the inspiration of a visual, operational knowledge (inspired by Alan Kay and Gilbert Simondon) which it finds in the concept of "diffraction" as it appears in and has been visualized for quantum mechanics. Diffraction describes the phenomenon of waves interfering with each other, although differences remain, much like in Thomas Young's image from 1803 (Figure 2.2) showing a two-slit diffraction.

The double-slit experiment with two waves interfering has become the thought experiment that is expressing puzzles of quantum mechanics, such as the wave–particle duality. In this century, diffraction also resurfaced as an interesting concept to think difference and was explored in depth in the writings of Karen Barad.[3] Inspired by particle diffraction of quantum trajectories, such as

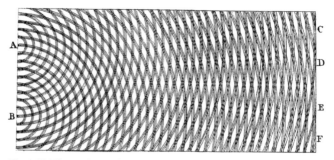

[Figure 2.2.] Thomas Young's sketch of two-slit diffraction presented to the Royal Society in 1803.

diffracted light waves, the philosopher with a doctorate in quantum
physics developed the method of reading of insights through one
another that came to be known as the *method of diffraction*. Barad
(2007, 137) is interested in the phenomenon of diffraction as it
allows her to think differences not as essentials but as a process.
Diffractive patterns are always fundamentally linked to the agential
apparatus that produces them, and vice versa: "Changing patterns
of difference are neither pure cause nor pure effect; indeed, they
are that which effects, or rather enacts, a causal structure, differentiating cause and effect." Here I'd like to take up Barad's aim of
deessentalizing difference but to mirror and link it to the difficulties
in differentiating the two modes in infantilization, that is, to be
empowering and patronizing at the same time. The circumstance
of infantilization's two effects—empowering and patronizing—
resembles diffraction: two waves that overlap to build a diffractive
pattern. The particles/waves overlap while the waves still can be
differentiated. Thus, as the image shows, despite them overlapping, there can still be difference. Or in other words, a diffractive
pattern, as we find it within the phenomenon of infantilization,
does not mean its effects cannot be differentiated. Following Barad
further, we therefore ask the question again: how can one conceive
this overlapping difference?

As Barad stresses, to understand diffraction, to know what kind
of diffraction is the case, it is important to look *further* than just
noticing that there is a pattern: "Crucially, diffraction effects are attentive to fine detail" (91). It is here where we find an aspect central
to her approach: the detail. In her own words: "Attention to fine
details is a crucial element of this methodology" (92). One has to be
"sufficiently attentive to the details" and is "thinking through the details" (73), because "fine-grained details matter" (90). It is the "level
of detail" (42) that enables one to answer a question. Thus it is to
the detail she looks to situate difference: "Small details can make
profound differences" (92). While the interference of the waves is a
given—otherwise, there would be no diffraction—the way a diffraction pattern looks can vary as it is linked to its parameters: "If any

of these parameters is changed, the pattern can be significantly different" (91). Only when looking at the details of the pattern and studying the "concrete" effects does one understand what exactly has been produced and which tendency of both—empowering or patronizing—precedes.

Unsurprisingly, pointing out those ambiguities and exploring their details also has become a habit of media and technology scholars interested in describing social formations. For this, theorists of digital technology and media have questioned word pairs like public–private, global–local, free–controlled, nature–technology, and work–play. Once understood as antithetical, they have made clear that their conceptual relation does not seem to be essentially oppositional anymore. Tiziana Terranova (2004) was among the first to discuss the ambiguity of work–play, pointing out that commenting online on platforms is free labor playing in the hands of companies looking for profit, although it remains pleasurable—a paradox. Wendy Chun (2011) also showed early that digital media is spreading democratic freedom along with the fact that it also accelerates the potential for global surveillance—an observation she later extended into digital media entering our daily habits, thereby messing "with the distinction between publicity and privacy, gossip and political speech, surveillance and entertainment, intimacy and work, hype and reality" (Chun 2016, ix). Analyzing algorithmic security practices and data technologies, Claudia Aradau and Tobias Blanke (2018) have disclosed how the dichotomies of normality–abnormality, friend–enemy, and identity–difference have been fundamentally reconfigured. Looking at the matter of media, Jussi Parikka (2015) dissects the opposition of nature–technology, which brings out the dependency of today's media from nature (Parikka 2015). Traversing computer science with a philosophical perspective, Luciana Parisi (2015) has questioned today's critique of instrumental rationality, pointing out that incomputability and randomness need to be conceived as the very condition of computation and not instrumentality. Pointing out dependence in a networked age, Anna Watkins Fisher (2016) discusses interventions

of corporations like Walmart or McDonalds, which aimed to help
their employees master problems created through being exploited
by the very same corporations. One could add Nicole Starosielski
(2015), Christopher Kelty (2012), N. Katherine Hayles's (2017) study
of the cognitive nonconscious, and many more whose recent
books or essays discuss how to deal with the ambiguities of new
media and the paradoxes we live with—the force digital technology
confronts us with.

These examples show that digital technology in the twenty-first
century is characterized by a dialectical setting in which disparate
aspects no longer operate in an oppositional mode, although their
dialectical relation has not collapsed—one is the flip side of the
other. Such a setting, in relation to the work of Pheng Cheah (2010),
could be described as "nondialectical dialectics." *Nondialectical*
as an interface that is addressing us as a very young child is both
patronizing and empowering and *dialectic,* as both moments are
still marked by an antagonistic relation, with one enabling the use
of power while the other is just lending it. Thus, regarding digital
technology, the task we face is to understand how to adjust the
frame in a way that fortifies the waves of empowering by turning to
the fine details. It is not to choose the right side.

⁞⁞⁞⁞⁞⁞⁞⁞⁞⁞⁞⁞

This chapter set out to study a force and found it linked to a figure
of power that it described as "nondialectical dialectics." Interested
in understanding how technology is addressing us, it aimed to
explore how a specific force unfolds in digital communication.
Drawing on Althusser's theory of interpellation, it identified a
particular situation opening up when being addressed by digital
technology communicating with us: digital interfaces, which aim to
reach a general user, show a tendency of infantilization. By drawing
on design elements from a child's world, such as big typography,
primary colors, big buttons, and animated mascots, those interfaces are addressing their users as young children, thereby calling
upon an experimental–operational knowledge rather than an
encyclopedic–scientific one. This type of knowledge, as could be

shown, has also historically been at the core of the development
of graphical user interfaces, which Alan Kay or Samuel Papert
conceptualized and built, inspired by the educational research of
Jean Piaget, who believed that the coordination of actions ordering
and joining things together should also be understood as "logical–
mathematical experience."

In this operational dialogue with digital technology, however, a new
phenomenon could be seen: it is not in a strict sense defined by a
dialectical logic of right or wrong dialogues with technology—and
in this lies the political sticking point. An interface that invites us to
an experimental dialogue exploring it can be empowering, while
it is not far from an interface that simply suggests how to use it
best without the user gaining any deeper knowledge about it (but
getting things done quickly). In other words, advising interfaces
that address us as children *can* but *do not have to* be empowering—
the force of digital technology that came into view could and
does go both ways. The cases analyzed here, from historic Google
Doodles to flat, colorful buttons on touch screens, are examples of
infantilization that show that the way digital technology is addressing us is deeply ambiguous. Digital technology can produce two or
more antagonistic effects at the same time and can therefore be
described as being nondialectical. Still, a dialectic relation remains,
as the effects it produces can be considered antagonistic with one
being the flip side of the other. Only when turning to the details
(Barad 2007), only when analyzing the actual effects, can the actual
political scale be understood.

The force of communication that then comes into view is a complicated, ambiguous one. It is a challenge—a challenge because
it is nondialectical while producing political effects; a challenge
because it has agency but is not an acting subject. When thinking
the force of digital technology, it helps to avoid understanding it in
an anthropomorphic way and to instead call upon its alien logic. So
I end this text with seconding what Finn Brunton pointed out in the
first chapter, who was preparing us for an alien dialogue in which
we find ourselves always already.

Notes

Without Wendy Chun's invitation and feedback on this contribution to, first, the Terms of Media II conference at Brown University and then to this volume, this text would not exist. Indeed, the text owes a lot to her encouragement here (and in other situations). I also owe warm thanks to the inspiration I got from the work and conversations with Finn Brunton and his aliens, waving to us through his text if one squints a little. Special thanks then go out to Paula Bialski, Goetz Bachmann, and Boris Traue for their thoughtful, informed, and thorough editorial reading of the manuscript, which improved it significantly. And thanks to the gifted Robert Ochshorn for sharing my serious interest in interfaces. Finally, I thank Michael Dieter and David Berry, whose invitation to contribute to their 2015 reader *Postdigital Aesthetics: Art, Computation and Design* (2015) gave me a first chance to grasp the idea of infantilization of digital interfaces. I am still surprised to find them sharing my perspective, the first time I presented it, which was the start that allowed me to build on it.

1   Understanding technology as a subject seems to be a projection linked to Finn Brunton's observation that human communication with aliens in space is imagined along the lines of a nonhuman agency with which we are familiar.

2   The analysis did not count individuals. Every time a new or a different sequence was introduced, it looked if the focus was on "adult" or "child," whereby groups counted the same as individuals. Three scenes were mixed. When the child plays football surrounded by a group of adults, the focus is mainly on the child (counted as child). The child birth in the surgery theater shows first adults at work; from there the camera moves to the child who was just born (counted as adult and child). The last scene shows Steve Gleason looking at the son on his lap (counted as adult and child).

3   Interestingly, Barad's strong focus on "interference" observed in the phenomenon of diffraction is somewhat close to Gilbert Simondon's approach, whose focus on the "ensemble" of technology and human—their interference—was discussed by describing the "technical reality" as one (Simondon 2017, 53). It has often been said (e.g., Combes 2013, 57) that Simondon's description of technology as an interference is informed by his concept of "individuation," which describes the process that produces an individual, although this individual is only a temporary instability—a theory he develops among others inspired by quantum and wave mechanics (Simondon 1992, 304), much like Barad. Therefore it comes as no surprise that Barad, with a doctorate in quantum physics, starts her point of departure—the preface of her book—from a very similar point of view. She writes, "Individuals do not preexist their interactions; rather, individuals emerge through and as part of their entangled intra-relating." Furthermore, she points out, "existence is not an individual affair" (Barad 2007, ix).

References

Althusser, Louis. 2014. *On the Reproduction of Capitalism: Ideology and Ideological State Apparatuses.* London: Verso Books.

Aradau, Claudia, and Tobias Blanke. 2018. "Governing Others: Anomaly and the Algorithmic Subject of Security." *European Journal of International Security* 3, no. 1: 1–21.

Arendt, Hannah. (1958) 1998. *The Human Condition.* Chicago: University of Chicago Press.

Barad, Karen. 2007. *Meeting the Universe Halfway: Quantum Physics and the Entanglement of Matter and Meaning.* Durham, N.C.: Duke University Press.

Bilton, Nick. 2014. "Facebook's New Privacy Mascot: The Zuckasaurus." *New York Times,* May 23. Accessed April 10, 2018. http://bits.blogs.nytimes.com/2014/05/22/a-blue-dinosaur-becomes-a-facebook-ambassador-for-1-28-billion-people/.

Bratton, Benjamin H. 2016. *The Stack: On Software and Sovereignty.* Cambridge, Mass.: MIT Press.

Brown, Stephen, and Sharon Ponsonby-McCabe. 2014. *Brand Mascots and Other Marketing Animals.* New York: Routledge.

Bunz, Mercedes. 2015. "School Will Never End: On Infantilization in Digital Environments Amplifying Empowerment or Propagating Stupidity?" In *Postdigital Aesthetics: Art, Computation, and Design,* edited by D. Berry and M. Dieter, 191–203. Basingstoke, U.K.: Palgrave Macmillan.

Bunz, Mercedes, and Graham Meikle. 2018. "Speaking Things." In *The Internet of Things,* 45–67. Cambridge: Polity Press.

Carter, Shan, and Michael Niessen. 2017. "Using Artificial Intelligence to Augment Human Intelligence." *Distill,* December 4. Accessed April 10, 2018. https://distill.pub/2017/aia/.

Cheah, Pheng. 2010. "Non-dialectical Materialism." In *New Materialism: Ontology, Agency, and Politics,* edited by Diana Coole and Samantha Frost, 70–91. Durham, N.C.: Duke University Press.

Chun, Wendy Hui Kyong. 2004. "On Software or the Persistence of Visual Knowledge." *Grey Room* 18: 26–51.

Chun, Wendy Hui Kyong. 2008. "On 'Sourcery,' or Code as Fetish." *Configurations* 16, no. 3: 299–324.

Chun, Wendy Hui Kyong. 2011. *Programmed Visions: Software and Memory.* Cambridge, Mass.: MIT Press.

Chun, Wendy Hui Kyong. 2016. *Updating to Remain the Same: Habitual New Media.* Cambridge, Mass.: MIT Press.

Coole, Diana. 2002. *Negativity and Politics: Dionysus and Dialectics from Kant to Poststructuralism.* New York: Routledge.

Combes, Muriel. 2013. *Gilbert Simondon and the Philosophy of the Individual.* Cambridge, Mass.: MIT Press.

Court of Justice of the European Union. 2014. Judgment in Case C-131/12 Google Spain SL, Google Inc. v Agenda Espanola de Protección de Datos, Mario Costeja Gonzalez. Press Release No. 70/14. Luxembourg, May 13.

Crawford, Kate, and Vladan Joler. 2018. "Anatomy of an AI System: The Amazon Echo as an Anatomical Map of Human Labor, Data and Planetary Resources." AI Now Institute and Share Lab, September 17. https://anatomyof.ai/.

Derrida, Jacques. (1977) 1988. "Signature Event Context." Translated by Samuel Weber and Jeffrey Mehlman. In *Limited Inc.,* 1–23. Evanston, Ill.: Northwestern University Press.

Economist. 2011. "Difference Engine: The iPad's Third Coming." December 2. https://www.economist.com/babbage/2011/12/02/difference-engine-the-ipads-third-coming.

Galloway, Alexander R. 2012. *The Interface Effect.* Cambridge: Polity Press.

Gitelman, Lisa. 2013. *Raw Data Is an Oxymoron.* Cambridge, Mass.: MIT Press.

Google Doodles Archive. 2018. "About." https://www.google.com/doodles/about.

Googleblog. 2010. "The Google Design, Turned Up a Notch." May 5. Accessed April 10, 2018. https://googleblog.blogspot.co.uk/2010/05/google-design-turned-up-notch.html.

Haraway, Donna. 1991. "A Cyborg Manifesto: Science, Technology, and Socialist-Feminism in the Late Twentieth Century." In *Simians, Cyborgs, and Women: The Reinvention of Nature,* 149–81. New York: Routledge.

Haraway, Donna. 1997. *Modest_Witness@ Second_Millennium.FemaleMan_Meets_Onco-Mouse: Feminism and Technoscience.* New York: Routledge.

Hayles, N. Katherine. 2017. *Unthought: The Power of the Cognitive Nonconscious.* Chicago: University of Chicago Press.

Heidegger, Martin. (1954) 1977. "The Question Concerning Technology." Translated by William Lovitt. In *"The Question Concerning Technology" and Other Essays,* 3–35. New York: Harper and Row.

Holt, Douglas. 2004. *How Brands Become Icons: The Principles of Cultural Branding.* Boston: Harvard Business School Press.

Iliades, Andrew. 2015. "Two Examples of Concretization." *Platform: Journal of Media and Communication Volume* 6: 86–95.

Intacto. 2013. Homepage. Accessed April 10, 2018. http://www.flatvsrealism.com/.

Kay, Alan. 1972. "A Personal Computer for Children of All Ages." Paper presented at the ACM national conference, Boston, August.

Kay, Alan. 1996. "The Early History of Smalltalk." In *History of Programming Languages—II,* edited by Thomas J. Bergin and Richard G. Gibson, 511–98. New York: ACM Press.

Kelty, Christopher. 2012. "From Participation to Power." In *The Participatory Cultures Handbook,* edited by Aaron Delwiche and Jennifer Jacobs Henderson, 22–31. New York: Routledge.

Kittler, Friedrich. 1990. *Discourse Networks 1800–1900.* Stanford, Calif.: Stanford University Press.

Kunzru, Hari. 1997. "You Are Cyborg." *Wired* 5, no. 2: 1–7. Accessed April 10, 2018. http://www.wired.com/1997/02/ffharaway/.

Langdon, Winner. 1986. *The Whale and the Reactor: A Search for Limits in an Age of High Technology.* Chicago: University of Chicago Press.

Levine, Rick, Christopher Locke, David Searles, and David Weinberger. 2000. *The Cluetrain Manifesto: The End of Business as Usual.* New York: Basic Books.

Mackenzie, Adrian. 2017. *Machine Learners: Archaeology of a Data Practice.* Cambridge, Mass.: MIT Press.

Marcuse, Herbert. 1998. "Some Social Implications of Modern Technology." In *Technology, War, Fascism,* 39–65. New York: Routledge.

Minsky, Marvin, and Seymour Papert. 1969. *Perceptrons: An Introduction to Computational Geometry.* Cambridge, Mass.: MIT Press.

Millman, Debbie. 2012. *Brand Bible: The Complete Guide to Building, Designing, and Sustaining Brands.* Beverly, Mass.: Rockport.

Noys, Benjamin. 2010. *Persistence of the Negative: A Critique of Contemporary Continental Theory.* Edinburgh: Edinburgh University Press.

Parikka, Jussi. 2015. *A Geology of Media.* Minneapolis: University of Minnesota Press.

Parisi, Luciana. 2015. "Instrumental Reason, Algorithmic Capitalism, and the Incomputable." In *Alleys of Your Mind: Augmented Realities and Its Traumas,* edited by Matteo Pasquinelli, 125–37. Lüneburg, Germany: meson press.

Papert, Seymour. 1963. "Études comparee de l'intelligence chez l'enfant et che le robot." *Études d'Épistemologie Génétique* 15: 131–94.

Papert, Seymour, with Guy Cellérier and Gilbert Voyat. 1968. *Cybernétique et Épistémologie.* Études d'Épistemologie Génétique 22. Paris: Presses universitaires de France.

Piaget, Jean, and Bäbel Inhelder. 1969. *The Psychology of the Child.* London: Routledge and Kegan Paul.

Piaget, Jean. 1972. "Development and Learning." In *Piaget Rediscovered,* edited by Richard Ripple and Verne Rockcastle, 7–20. Ithaca, N.Y.: Cornell University Press.

Raley, Rita. 2016. "Algorithmic Translations." *New Centennial Review* 16, no. 1: 115–37.

Rawsthorn, Alice. 2010. "Google's Doodles." *New York Times,* March 24. Accessed April 10, 2018. http://www.nytimes.com/2010/03/14/t-magazine/02talk-rawsthorn.html.

Shannon, Claude, and Warren Weaver. 1949. *The Mathematical Theory of Communication.* Chicago: University of Chicago Press.

Shultz, Thomas R., William C. Schmidt, David Buckingham, and Denis Mareschal. 2008. "Modeling Cognitive Development with a Generative Connectionist Algorithm." In *Developing Cognitive Competence: New Approaches to Process Modeling,* edited by Tony J. Simon and Graeme S. Halford, 205–61. Hove: Lawrence Erlbaum.

Simondon, Gilbert. 2017. *On the Mode of Existence of Technical Objects.* Minneapolis, Minn.: Univocal Press.

Simondon, Gilbert. 1992. "The Genesis of the Individual." In *Incorporations,* edited by Jonathan Crary and Sanford Kwinter, 297–319. New York: Zone Books.

Simondon, Gilbert. 2012. "Technical Mentality." In *Gilbert Simondon: Being and Technology,* edited by Arne Boever, Alex Murray, Jon Roffe, and Ashley Woodward, 1–15. Edinburgh: Edinburgh University Press.

Starosielski, Nicole. 2015. *The Undersea Network.* Durham, N.C.: Duke University Press.

Terranova, Tiziana. 2004. *Network Culture: Politics for the Information Age.* London: Pluto Press.

Tesco PLC. 2015. "New Audio Voice for Self Service Checkouts." Accessed April 10, 2018. https://www.tescoplc.com/news/news-releases/2015/tesco-to-end-unexpected-item-in-the-bagging-area.

U.S. Naval Observatory. 2018. "USNO GPS Timing Observations." Accessed April 10, 2018. http://tycho.usno.navy.mil/gps.html.

Watkins Fisher, Anna. 2016. "User Be Used: Leveraging the Play in the System." In *New Media, Old Media,* 2nd ed., edited by Wendy Hui Kyong Chun, Thomas Keenan, and Anna Watkins Fisher, 287–300. New York: Routledge.

Williams, Raymond. 1985. *Keywords: A Vocabulary of Culture and Society.* Oxford: Oxford University Press.

Wingfield, Nick. 2012. "Microsoft Drops Metro Name for New Product Look." *New York Times,* August 3. Accessed April 10, 2018. http://bits.blogs.nytimes.com/2012/08/03/microsoft-drops-metro-name-for-new-product-look.

Winner, Langdon. 1989. *The Whale and the Reactor: A Search for Limits in an Age of High Technology.* Chicago: University of Chicago Press.

3

Code Review as Communication: The Case of Corporate Software Developers

Paula Bialski

Communication is a rich tangle of intellectual and cultural strands that encodes our time's confrontations with itself. To understand communication is to understand much more.

—John Durham Peters

In early August 2017, a Google employee published an internal memo titled "Google's Ideological Echo Chamber." Most of us recall the content of the memo as well as the buzz that it stirred in tech circles in Silicon Valley and beyond: the memo outlined the reasons that Google's diversity initiative was fundamentally wrong. One section in particular struck me—an ethnographer of corporate software developers—as particularly misleading: "as Google was trying to make software engineering more people-oriented with pair programming and more collaboration," Google's initiative was fundamentally limited by "how people-oriented certain roles and Google can be" (Motherboard, August 8, 2017). One of the memo's messages was to "deemphasize empathy" in tech companies. As far as software developers are concerned, who needs empathy anyway when building technical systems?

While I acknowledge that this memo stirred up a lengthy debate around gender biases in the tech industry, about what women are good and not good at, and about the general differences between men and women in the first place, I will, for the purpose of staying on topic, leave out this debate and focus on one of the article's claims: that communication and empathy should not be stressed as assets among tech workers.

Communication in digital cultures is multifaceted. We know that it can exist between humans, mediated by digital technologies. It can occur between a human and machines (the way we record our voices or type into a screen). It can exist machine to machine (the way Finn Brunton so eloquently underlined in chapter 1 of this volume).

What the author of the Google memo overlooked was that software is perhaps one of the most complex examples where digital media and human sociality indeed intertwine. Human empathy in a team meeting sits alongside merging pieces of code written by separate people. In this chapter, I would like to zoom in on the processes of a large-scale software development corporation to uncover the way in which human–machine cohabitation and co-creation call on various modes of communication. In previous chapters, Brunton explored communication and digital technology through questioning the alienlike dialogue between technical entities—focusing more on communication between machines—and Bunz looked at the way in which these technical entities are designed to address us, the user, and what kind of effect this type of design (childlike and patronizing) has on us. This chapter will draw you, the reader, into an ethnographic field to see the way communication between machines, between humans and machines, and between humans building machines plays out. Doing so, I hope, will give you a grounded example of the multifaceted nature of communication in digital cultures.

Charlie

I met Charlie about two years ago during one of Noah's film nights in Berlin's Neukolln district. We were all sitting on Noah's living room couch and discussing their software company—a place where I later started conducting my long-term fieldwork project. The company focused on building mapping and navigation software and had approximately six thousand employees. One thousand of them were software developers of all shapes and sizes. I worked with Charlie on the "navigation" team, which worked on, among other things, how to find the best route from A to B (which is a much tougher problem than it sounds).

I just assumed Charlie was a developer. And he played the part. He spoke knowledgeably about code, bugs, breakdowns, and updates, and I somehow just jumped to the conclusion that he knew a lot about programming. It was only in the recent weeks, when I started working in his department, that I noticed something fishy. During a product owner[1] meeting, he was there. Sitting at the table. He was very engaged in the meeting, very vocal, very un-developer-like. I first figured that he was a lead developer, or somebody just keen enough to sit at the product owner meetings, until I realized that he was, in fact, a product owner. This mix-up—the fact that I didn't know what an employee was doing exactly—was quite a common mix-up, even for any regular employee. This was due to the fluidity of both his role (he himself didn't know if he was a product manager, product owner, error manager, or a number of other possible positions) and the amount of technical knowledge he had (his knowledge of the software system they were building was somewhere in the gray zone of being technical and not-so-technical-at-all).

Charlie had the poshest of English accents. This accent would surprise me every time he opened his mouth. Perhaps it was because this software development company hires a mishmash of nationalities, creating a sort of Globish clang around the office. To overhear a real native English speaker was quite odd. And the type of English he spoke was a Merry Poppins sort of English, using sentences like "I'm just taking care of some bits and bobs" and "we're far away from everything being tickety boo." Despite the fact that he looked cheery all the time, Charlie always explained that he was stressed and that there was a lot of work on his shoulders.

Charlie was also very understanding of my role as an ethnographer and spoke to me softly and with true interest in what I do. That afternoon, after I asked him a few times about the "fire" he was trying to put out, he offered to show me what he was working on. He grabbed a coffee cup and his laptop, and we walked over to the common area near the staircase, with nice wooden picnic tables. He opened up his laptop to show a screen of software called Gerrit—a free, open source, web-based code collaboration tool.[2] Gerrit displayed a list of *merges*—meaning updates or additions to the code base. This list included the type of update, who updated it, and the status of the update (whether it passed or failed). During merging, the code would get automatically tested, and various problems could arise. I learned that it was his job as error manager (an additional job he was assigned to do, on top of his product owner job) to monitor these updates, especially during the last week of production. That day was *feature complete day*—which meant that all features or changes to the new software version should be finished and merged into the code base. Merging doesn't happen seamlessly. Charlie anticipated that something wrong would happen and, during days like these, always kept his computer open and monitored the progress of each merge.

Looking over his laptop, he started pointing at a list of bars colored red and light blue, delineating the status of the merge process. Each of these bars was one "piece" of code belonging to one developer who pushed this code into the rest of the build. The bars would change color depending on whether the code merged, or worked, with the rest of the code. I looked at one of the bars that turned red and said "merge error," and I asked him, "So is it kind of like putting two pieces of a puzzle together, but one person chopped off one arm of the puzzle piece, so it doesn't fit into the other piece as it was intended to?"

Charlie explained, "You have to imagine that there is a Google Doc that two people are working on. And at the same time, one editor makes changes to what the other editor was working on. When they merge these two documents, their changes would conflict."

Charlie explained that this is called a *merge conflict*—when a change, or *job,* that one developer merges influences another line of code that another developer merged. We were both looking intently at the screen at a long list of around fifteen jobs, some of them with "merge conflict" marked next to them. Throughout the next fifteen minutes of sitting there together, I could see him monitoring this screen intently. He explained that each time a developer wants to merge what he or she was working on into the main repository, it would take around forty minutes for the server to test the merged code. In monitoring this screen, Charlie was waiting for a developer, and the server, and the code base to communicate something to the rest of the team (including him) regarding the status of the code. And there were also other conflicts that could arise aside from merge conflicts.

I noticed myself starting to get a small sense of how programmers worked together. And the way they worked together was through various forms of communication, avoidance of communication, or analyzing communication. The term *communication* was constantly floating around the office. If we take Charlie as an example, just these few minutes of interacting with him displayed all the verbal, nonverbal, machine-based, human–machine, or team-based communication that takes place in a software development setting. Charlie's type of English compared to the English of his colleagues posed a communication issue. The way Charlie's client expressed what they needed Charlie's feature to do for the user posed a communication issue. The way one developer's code communicated to the code base posed a communication issue. The various ways two developers worked on the same piece of software posed a communication issue.

Moments of Unfolding

The question of how digital media technology affects the social and how the social affects digital media technologies can best be studied in such hard-to-grasp, always slippery, always changing moments of unfolding. My goal with this initial fieldwork anecdote was to briefly introduce you to the highly complex feat of communication that software development entails. Mixed in between servers; shyness; Globish; and software standards, testing standards, and customer expectations are people collaborating on the same project. This collaboration consists of a deep intermingling of human and machine. As Mercedes Bunz pointed out in her chapter in this volume, digital communication needed to be conceptualized on more than one layer, because software operates always with more than one interface. These interfaces are between machine and machine and between machine and user. As I will point out, there is also another layer, that between the writers of code (software developers) and other software developers. The overarching goal in this chapter is to speak to the issue of communication in digital cultures, by providing a rather descriptive map of the forms of communication that occur in corporate programming environments. In doing so, I aim to explain exactly how software development is organized as a practice and how communication fits into this practice. This communication that I will unfold includes a number of objects and people. This communication takes place

1. machine to machine
2. programmer to machine
3. programmer to organization
4. programmer to programmer

As part of a constant assemblage of human and machine, these three sides—organization, producers, and the material resistance of software—interact with one another in iterative processes, shaping together digital media, which are never not just filters but, in John Durham Peters's (2015, 2) words, "vehicles that carry and communicate meaning." The methodological challenge is then in understanding how all these pieces communicate with each other and to observe such relations in moments of their unfolding.

In this chapter, I mainly draw on the ethnographic data I collected during my ongoing research of corporate software developers. In August 2016, I began an organizational ethnography at a large, six-thousand-employee Berlin-based software company, BerlinTech. At the time this chapter was published, I had spent more than eighty working days immersed in this company—attending meetings, observing the developers develop software, and holding both formal and informal interviews in the office kitchens, over lunch, or during long walks to and from work. My general focus within my research is to uncover technologically mediated sociality. I frame my research within cultural, social and media theory, science and technology studies, software studies, infrastructure studies, and the ethnography of software developers.

As I mentioned in a previous paragraph, this company makes key mapping infrastructure for third parties such as social networking sites or various apps in need of embedded maps. I spent two months among front-end app developers who built the company's "consumer" product—meaning the app that users download for use on their smartphones and desktops—and am currently situated among the back-end developers who build in-car navigation systems and maintain the routing infrastructure. The larger motivation behind my ethnographic project stems out of the belief that studying the mundane, quite "average" software developer brings us closer to the way in which our digital infrastructures work. To study everyday software developers—the masses of engineers who build the infrastructures that power our mobility, our sociality, our consumption (the list goes on)—is to study of how communication between humans and machines functions, and is changing, in our everyday, digitally networked lives.

In this chapter, I attempt to unpack how developers communicate with one another, and how machines communicate to developers, as well as the imperfect tools (such as the code review software) that are invented to assist the communication between machines (code to code).

This research aims for (1) an ethnographically rich understanding of digital sociality that (2) focuses on moments of unfolding; (3) is based on a trimodal way of communicating between developers, the code they build, and the software they use to communicate with; and (4) last, but not least, looks at the interrelations between these three sides. I am aware, of course, that my field has other sites where communication takes place, especially between designers, management, and users. Programmers use a variety of collaborative management tools that are designed specifically for developers themselves (such as code review systems and workflow systems) or for the labor force at large (such as electronic calendars, intranet systems, and online spreadsheets). For the sake of this topic, I narrow down our case study of developer communication to one tool, Gerrit, that strictly relates to a practice called *code review.*

Despite ethnographies that look at the intricate ways in which programmers collaborate, an ethnographic mapping of programming practices in corporations in relation to the notion of communication has not been done. Many of these topics have also been addressed extensively in software engineering literature, but not from an ethnographic angle.

Describing Gerrit

After two months of working in the field, my field diaries left me with lingering questions. The luck of my field was that my informants weren't an isolated tribe, cut off from the rest of the world, but were software developers, who, as you could imagine, were often sitting in front of their computers, often with their various chat systems (Skype, Spark, or Slack) open, instantly accessible. This provided me with a sort of never-ending feeling of field access—as if I had a lifeline to my field that hardly ended after I left the office.

That being said, this chapter is made up of a variety of pieces of data I collected throughout a two-year period: a number of field notes before I entered the field, notes collected during my fieldwork, and more recent online exchanges with my informants over various chat applications.

The conversation between Charlie and me illustrated the basics of a code review system: it is based on a culture of peer review. The team of one hundred developers I currently research creates routing infrastructure for navigation systems (so how the fastest or most optimal route from A to B is calculated). These developers are then split up into subteams. The team I directly sit with is the electric vehicle (EV) team, comprising five developers. They collectively have to solve a routing problem typical for EVs: how to optimize a driving route for a car that has to secure charging stations every few hours? Before their project is finalized and is implemented into an in-vehicle system, their work is split into small subtasks. Before the project starts (the project's *sprint period*), the group of developers sit together in a room with their manager, or *product owner*, and define the subtasks that need to be done to complete the project. These subtasks are called *tickets*. Then, the next day, each developer takes a ticket and starts working on it. A ticket, in the case of the EV team, could be, for example, to match the library for EV charging stations to a route library. Maria would then take this ticket, work on it, and then upload it or "push" it to the Gerrit code review system for review. Her colleagues would then give her a score between +2 and −2 (the exact meaning of which I shall explain later).

Noah, one of my main informants, explained that the main purpose of the code review[3] was to collectively monitor if each line of code that a developer tries to upload into his team's main repository actually fits, and will work, with the code that the other team members have created. There are two main threads of communication at play here: the communication between developers, both offline (e.g., developers often peek over their desktop screens and yell out to a colleague, "Can you review my code please?" or "Hey, why did you give me a −1!") and via their code review software (i.e., in the form of a code review), and the communication that lies between the technical system itself, for example, Noah's lines of code have to speak to another developer's lines of code, and these collections of lines have to speak to one another, so to speak, when running within the entire system. While I will get into more detail in a later section, I just want to underline that in this process, communication in the right way, using the right syntax, for example, is highly necessary.

After being a year in this field (with only latent knowledge of how software is built), I have come to the understanding that code review systems are an inherent part of a corporate software environment—a crucial part of a sort of production pipeline. It is necessary because a software product has to be shipped to its users within a certain time frame for the software company to remain competitive on the market, and that shipment has to actually work, without major bugs or breakdowns. Large-scale corporate software environments, made up of teams of dozens of developers, also need core review systems to make collaboration (and communication) with a large number of software updates easier. Michael, a developer who used to work at BerlinTech and who has since moved to a small start-up company, explained that the code review system at BerlinTech is a good example of how a big company deals with the review process. After moving jobs, Michael and his team of two to three people don't have a review process, but rather, the team communicates with one another by talking before starting their job, agreeing on something, and cooperating throughout the development process. Merging a change into the whole system becomes more of a formality in his case, rather than a necessity.

In noncorporate contexts, where software developers work on a small team without a strict deadline, they might use an informal code review system—giving each other feedback much like a band of musicians would give each other tips and tricks on how to make a song better. But a code review system in a corporate, large-scale software project environment is a formalized, software-based assembly line. Additionally, the code review system is a way of monitoring the code that goes into building a system but also becomes a surveillance tool for upper management to monitor what type of work the programmers are actually doing. And as I shall explain in a later section of this chapter, it also becomes a sousveillance tool for programmers to monitor other programmers.

At BerlinTech, Gerrit is only one of many code review systems in use among software developers. The company itself has more than one thousand developers, and communication between developers happens in a number of ways, including through code review systems, but also in face-to-face, daily, fifteen-minute standup meetings; at lunchtime; at other collaboration meetings; and via all sorts of internal (meaning set up by the company for office use only) and external (meaning set up by team members for office use and other issues) platforms. Some (and this list is in no way exhaustive) include the company's internal chat system, called Yammer (which is limited for social interaction), which is the company's "official" internal communication tool for all work-related issues, and collaboration software like Cisco Spark, Slack, or even Skype chat. Each team, depending on what they are building, has a different type of communication tool, and even a different type of code review system. Keeping track of these channels of communication was a dizzying and oftentimes frustrating task for the developers with whom I worked. Some collaboration tools emerge bottom-up, initiated by the collaborators themselves, and some collaboration tools were initiated top-down, by the management of the company. While the ways in which sociality emerges in these meetings—in specific working groups or on each individual platform, with their specificities and power politics—are relevant to the theme of communication here, I will refrain from addressing them for the sake of keeping a more narrow focus on code review tools and the communication inherent within them.

Returning to Gerrit, this is a code review system that is free and open source, originally authored by Google—and it is not the only system used at BerlinTech. As Noah explained, the reason many different teams use different code review systems is "mostly because not everybody completed the transition to Gerrit. Some people prefer other ways, because their project is small, or they're not aware of Gerrit, or they just don't like it" (field notes, June 2017).

Simon, one team leader, explained to me, "Gerrit is a key part of our culture. If a developer has a piece of code, he uploads it to Gerrit. You collaborate together to make one commit happen. This is not like competition for making code. It is trying to work together. Trying to transfer information and knowledge" (field notes, August 2016).

Sebastian, another iOS developer (building the application for iPhones), uses the metaphor of a tree to explain how code review works: "It's like a tree and every coder adds another branch to that tree. In order to merge their changes (branches), there has to be certain tests done. Only after these tests can the branches be really incorporated as part of this tree" (field notes, August 2016).

Using Sebastian's metaphor, when a programmer adds a "branch" to Gerrit, it is visible to other developers, and the code waits for at least two developers—plus an automated bot—to approve the code. At BerlinTech's front-end team, developers were encouraged to look at an incoming review every one to two hours, although one developer informed me that this rule "often didn't happen anyways, but a review of your work did take place every twenty-four hours" (field notes, June 2017). A review in the Gerrit system has five variables:

-2: Do not submit

-1: I would prefer that you didn't submit this

0: No score

+1: Looks good to me, but someone else must approve

+2: Looks good to me, approved

Michael, a web developer, half-jokingly once confessed that on Fri-
days, when he feels like leaving work and running off for a beer, he
would quickly go through the code review system and just add "+2,
+2, +2" to all the tickets waiting to be reviewed (field notes, August
2016). How much of this is actually true is a mystery, but it alludes
to the way in which variables such as fatigue, the weather, the time
of day, and the relationships between the developers themselves
factor into their ratings.

Communication Models

At this point you, the reader, are perhaps beginning to understand
the general purpose of a code review system and some basics
of how Gerrit functions. If we unpack how communication works
within this system, we can start to see three communication models at play. First, the code review system mediates communication
between people, in this case, programmers working on code. This
type of model of communication is between two humans and is
dialogic and interactive. In the case of Gerrit, developers review
one another based on a culture of reciprocity: if somebody has
reviewed your code, you are obliged to review her code in return.
If a developer requests a code review and then, upon gaining a
review, refuses to reciprocate and review code in the future, this
behavior is viewed as antisocial. Moreover, code review is also
done in conversation. For example, a developer might write, "To be
honest, this whole solution looks a bit . . . hacky. Could you please
describe how it works and why this is a correct solution?"[4] This
developer will not approve the "committer's" (programmer submit-
ting a code update/solution to be reviewed) code until he gains an
answer to his question. Code review is a continuous back-and-forth
between committer and reviewer and must reach a level of mutual
understanding.

Additionally, it might not come as a surprise that interaction
between two developers is often easier through a system like
Gerrit, minimizing the social pressure of face-to-face interaction.
Michal recalled a situation where "there was one review that had

20+ patches [fixes] and a lot of discussion in the comments just between two guys. And as I see it, it would be way easier for them to chat (one was working remotely) and decide on a solution and then just fix it with one patch. But they were both pretty introverted so I'm guessing they 'preferred' to handle it through Gerrit" (field notes, July 2017).

As Peters (1999, 23) has noted, this model of communication is also sometimes defined by "open and frank talk between intimates or coworkers. Here communication does not mean simply talk; it refers to a special kind of talk distinguished by intimacy and disclosure." Michael also added that in the best iteration of communication during code review, the two developers share a sort of "respect": when the reviewer asks questions about the solution, the committer attempts to answer them and fix the problems that the reviewer mentioned (field notes, June 2017). This respect can form an intimacy between two developers, where two developers form a sort of rapport through their code review process.

Communication in this case is thus reciprocal, takes place between two developers, and involves explicit discussion and questions that one developer directs toward another. The rating system I mentioned in an earlier section (+2, +1, 0, −1, −2) is also part of a more implicit communication system between developers, and each developer might give his own subjective understanding of what each score means without explicitly describing this meaning to the other developer. This type of communication also can mean a sense of "transfer or transmission . . . of ideas, thoughts, or meanings" (Peters 1999, 15).

I asked Michael, "OK, so let me get this straight. I can't give Mike a +2 if a little piece of his code is wrong, so then I comment on a code block, and then Mike patches it, and its still wrong, so I comment again?" Michael responded by stating,

> Well you can—it depends on the reviewer style; my style was, "I can give you +2 even if something is a bit wrong, as long as I will highlight it and you will fix it in the future,"

because my main goal with code review was to make the other person a better programmer for the future. But I know some asshole reviewers who would see a very small bad piece of code . . . for instance, a variable name they didn't agree with, because it wasn't descriptive enough, and they would give the committer −1 and tell them to fix it. I would in that case give them +2 and say "make the variable more descriptive in the future" . . . but I can imagine the other side would say "I care about the code-base and I'm strict with reviews, but some other people [like Michael] allow bad code to go through and assume that people will fix it." And a "lazy reviewer" I'm assuming would give a −2 to code without explaining what's wrong. (field notes, June 2017)

In this example, Gerrit's rating system takes on different standards for different developers. Where the so-called asshole developer gives a −1 to a developer who forgets to provide enough description of a variable, Michael chooses to waive this lack of description for the benefit of encouraging the developer to work on it next time. Technical standards take on different meanings in practice, and developers who understand the meaning of what Michael is communicating with his +2 will be all the more annoyed or surprised when they receive a −1 from him.

Broadcasting

The second model is the broadcast model of communication: rather than being a reciprocal exchange between dyads, code review becomes a one-to-many performative act. The developer's review itself broadcasts meaning to a group of (often) unspecified others.

For instance, Gerrit functions in such a way that each review is tagged with the name of the reviewer. At BerlinTech, one development team I was working on was expected to conduct at least two reviews per hour. While this standard was not kept up, some developers would just engage in code review to keep up their hourly quota. Note that neither the content of the code review comment nor the rating of the code review itself was important. Rather, the mere gesture communicated to others "I am reviewing, thus I am part of this team." If Noah, for example, conducted a review, he was communicating to his fellow developers, as well as to his supervisor, that he is indeed a team player and does care about the quality of the team's project.

On the other hand, I observed that some developers found code review to be part of their procrastination regime, as it was easy and "mindless." Engaging in code review would help communicate to their team as well as their boss that they were engaged and working, while in practice, they were merely conducting this "mindless" work to avoid their more challenging job. The reviewer is thus not engaged in a meaningful dialogue with the committer but is rather clicking through the reviews quickly and mindlessly.

As I also mentioned earlier, code review becomes a surveillance tool for upper management to monitor what type of work the programmers are actually doing. Gerrit is designed in such a way that makes work highly transparent to anyone in the company, including management. It also becomes a sousveillance tool—a way for colleagues to implicitly monitor one another's work. The mere online presence of a manager within the Gerrit system, as well as the mere presence of a colleague, broadcasts or communicates to other developers "I am here, and I am watching you."

Communicating Code

A third form of communication present within a code review system is communication between pieces of code, which are essentially lists of commands or solutions that communicate meaning to a machine of other commands or solutions. Code can be seen as a sort of agent, commanding the computer's performance.

My previous two definitions of communication involved human communication. But in the case of code review, code has to speak to other code for it to function. Peters (1999) noted that while *communication* can mean a sort of exchange, or transfer, involving some sort of reciprocity, these factors do not have to take place between humans: "Communication can mean something like the successful linkage of two separate termini, as they say in telegraphy. Here simply getting through, as in delivery of mail or e-mail, is enough to constitute communication" (16). In this book, my fellow author Finn Brunton eloquently analyzes Claude Shannon when discussing this type of communication between machines.

To illustrate how this functions in a code review system, take a look at a snippet of conversation between a team of developers who are discussing a small functionality failure in a feature they are developing for a website. This conversation was cut from a Skype conversation they were having (as mentioned, developers often discuss their issues using various chat systems):

NINA: There is one failure, which I think not caused by you, @John—"Scenario: Clicking on map markers and PDC links" . . . when I check in the commit of Michal I still see it failing.

JOHN: well, the current run is my submit. Let's see what is left after that, but yea, worth taking a look at as well

NINA: @Piotr, I see that the step "And I expand the place panel contents if not expanded" which you added. I wonder if the change you did is the reason of failure of "Scenario: Clicking on map markers and PDC links."

In this instance, John submitted a change (often called a *commit* or a *submit*) to the functionality of a side panel that is affecting various links and markers on a map on their website. In this instance, the code in John's commit is not communicating to the code in Piotr's submit. Perhaps this code has a syntax error—which could mean that the line of code is not merging or communicating with the other code through this error. Unlike humans, who can infer meaning even when something is mumbled or slurred, or when a word is spoken in a different language, when various parts of one machine speak to one another through code, this code has to be written using the correct syntax. One can say that the main purpose of the code review process is for humans (programmers) to assist machines (various parts of an app build) in their communication process.

Code communicates to other code through these processes of merging. Maria, the back-end developer I mentioned previously, explained the way in which code interacts, by stating, "Yes, we are constantly breaking each other's stuff. What you are creating communicates with other code others are building. And we are all doing it at the same time. Its like building a house. When building a house, the plumber works on something, and then the carpenter. Often this doesn't happen at the same time, because their work would conflict with one other. But we do work on stuff at the same time. And it does conflict."

One can imagine a sort of Tower of Babel, with code communicating to other layers of code. Although each line of code, or each commit, can be traced back to the author who created it, code has to communicate with other code even when developers are away, on vacation, or asleep. This communication brings the machine (in the case of my developers, a routing engine) to life.

‖‖‖‖‖‖‖‖‖‖‖‖

Technical systems are incredibly social. A code review system, based on something that should be seemingly mechanic, is a highly variable communicative process both because of its technical limitations and also because of the culture of communication that develops in the setting in which it is situated.

Understanding how developers communicate to one another through a code review system can reveal how a technical system structures cooperation, how standards of communication develop, and the idiosyncrasies of human–human as well as human–machine communication. It can help us understand how digital media at large impacts communication. Understanding how producers and designers of technology create their products can reveal the cultural, political, and economic motives that structure and determine the design of technology. Understanding the material resistance and viscosity of software, its bugs, breakdowns, and technical dependencies, helps us to understand some of the properties of digital media technology itself.

This chapter has focused on one case study: the code review process among corporate software developers. I did so to understand the various nuances and forms that communication in digital cultures can take. Although my communication models were in no way exhaustive here, I focused on "communication as interaction" between programmer and programmer, "communication as broadcast" between programmer and the rest of the organization, and "communication between code" as the way in which code is seen as an interactive process between other code. This analysis, I hope, brought you, the reader, closer to imagining the communicative interrelationship between machines (compilers, databases, processors, memory, servers, clouds), (2) programmers, and (3) the infrastructure within which both programmers and machines function.

Notes

1   Those less technical members of a development team who held the developers responsible to the customer and the product they were creating.
2   See http://www.gerritcodereview.com/.
3   As one of my informants clarified, the code review has many unofficial roles as well, in building shared coding practices, initiating junior developers, building a shared view of the team on how code should look, elucidating which part of code they are putting emphasis on as a group—style, efficiency, testability, and so on. In less collaborative teams, one can see how code review creates conflicts that should have resolved prior to coding.
4   https://github.com/facebook/react-native/pull/14259.

References

Peters, John Durham. 1999. *Speaking into the Air: A History of the Idea of Communication*. Chicago: University of Chicago Press.
Peters, John Durham. 2015. *The Marvelous Clouds: Toward a Philosophy of Elemental Media*. Chicago: University of Chicago Press.

Authors

Paula Bialski is junior professor for digital sociality at Leuphana University Lüneburg and author of *Becoming Intimately Mobile.*

Finn Brunton is assistant professor in media, culture, and communication at New York University. His most recent book is *Digital Cash: The Unknown History of the Anarchists, Utopians, and Technologists Who Built Cryptocurrency.*

Mercedes Bunz is senior lecturer in the Department of Digital Humanities, King's College London. Her last book, written with Graham Meikle, is *The Internet of Things.*